STUDIES IN
INTERNATIONAL RELATIONS

edited by
Charles G. MacDonald
Florida International University

A ROUTLEDGE SERIES

STUDIES IN INTERNATIONAL RELATIONS
CHARLES G. MACDONALD, *General Editor*

THE COMMON FISHERIES POLICY IN THE EUROPEAN UNION

A Study in Integrative and Distributive Bargaining

Eugénia da Conceição-Heldt

Routledge
Taylor & Francis Group

NEW YORK AND LONDON

Published in 2004 by
Routledge
711 Third Avenue,
New York, NY 10017

Published in Great Britain by
Routledge
2 Park Square, Milton Park,
Abingdon, Oxfordshire OX14 4RN

10 9 8 7 6 5 4 3 2 1

Library of Congress Cataloging-in-Publication Data

da Conceição-Heldt, Eugénia
    The common fisheries policy in the European Union : a study in integrative and distributive bargaining / Eugénia da Conceição-Heldt.
        p.   cm. — (Studies in international relations)
    Includes bibliographical references (p.) and index.
    ISBN 0–415–94902–5 (hardback : alk. paper)
    1. Fishery policy—European Union countries. 2. Negotiation—European Union countries. I. Title. II. Series: Studies in international relations (Routledge (Firm))
    SH254.E87C657 2003
    338.3'727'094—dc22

ISBN: 978-0-415-94902-6 (Hardback)

ISBN: 978-0-415-64894-3 (Paperback)

Contents

List of Figures and Tables

LIST OF FIGURES

LIST OF TABLES

List of Figures and Tables

Acknowledgments

In the process of writing my dissertation, on which this book is based, many people have helped, directly or indirectly. Trying to acknowledge the contributions of the people who supported and helped me in many ways during the last three years is a humbling task. At the same time, it gives me the pleasure of thanking those who guided my academic preparation and those who shared these times with me as friends and colleagues. First of all, this study would not have been possible without the diligent supervision and the intellectual stimulus of Prof. Dr. Michael Bolle, who transmitted to me his passion for rational-choice and public choice theory. I am also indebted to Prof. Dr. Otto Keck and to Prof. Dr. Thomas König for their numerous insights into the subject and constructive criticism. Many colleagues and friends have provided valuable encouragement and help to me during the course of writing this dissertation by reading and commenting on parts of the manuscript. I should like to express my gratitude in alphabetical order to: Uli Brückner, Christian Fahrholz, Hans-Walter Heldt, Achim Kemmerling, Charles MacDonald, Mark Pollack, Uli Sedelmaier, Sabine Schwarz, and Bernhard Zangl. I also wish to thank very much Fiona Heldt, who faithfully corrected endless versions of this manuscript. For the graphical assistance I thank Stephan Hohenberger. Despite the extensive intellectual contributions from many people, I alone take responsibility for what I have written.

I also wish to acknowledge publicly my large debt to two institutions which directly contributed with resources to this research project. My gratitude goes to *the Studienstiftung des deutschen Volkes* for providing me with financial support during the last two and half years. I am also grateful to the Jean Monnet Centre of Excellence of the Department of Political Science at the *Freie Universität Berlin*, and especially to Prof. Dr. Michael Bolle, for providing the academic space to research this topic.

Finally, this book is dedicated to Oliver Heldt for his support and for helping me see over the last years that there is more to life than what you are about to read.

Berlin, September 2003

Abbreviations

BATNA	Best Alternative to a Negotiated Agreement
CAP	Common Agricultural Policy of the European Community
CFP	Common Fisheries Policy of the European Community
COPA	*Comité des Organisations Professionelles Agricoles*
COREPER	*Comité des Représentants Permanents*, Committee of Permanent Representatives of Member States to the EU
EAC	European Affairs Committee of the Danish national parliament
EC	European Community
EEC	European Economic Community
EEZ	Exclusive Economic Zone
EP	European Parliament
EU	European Union
Euratom	European Atomic Energy Community
FAO	Food and Agricultural Organization of the United Nations
GATT	General Agreement of Tariffs and Trade
Ifremer	Institut français de recherche pour l'exploitation de la mer
MS	Member State
OECD	Organization for Economic Cooperation and Development
QMV	Qualified Majority Voting
SQ	Status quo
TACs	Total Allowable Catches
UNCLOS	United Nations Convention of the Law of the Sea

Introduction
Analyzing negotiations in the European Union

The recent bifurcation of European studies into state-centric and new institutionalist camps has resulted in a sterile theoretical debate that says little about an empirical world where bargaining outcomes cover both member states' preferences and institutions. This work is an attempt to move beyond the theoretical debate. It elaborates on the gap between theoretical and empirical studies. For this a conceptual framework for analyzing negotiations in the European Union (EU)[1] will be juxtaposed to the key negotiations leading to the establishment of the Common Fisheries Policy (hereinafter referred to as CFP). It is a stock phrase that national governments' preferences matter in the EU, but what is less precisely known is whether national preferences are really central in the negotiation process or only one variable among many others. Nevertheless, "taking preferences seriously" (Moravcsik 1997a) is not the whole story of bargaining in the EU. In the end member states make the ultimate decisions, but they do not fully control the negotiation process. Member states are constrained by the preferences of the European Commission[2], considered to be an actor in its own right, and also by the institutional setting, which includes the variables limiting the range of feasible alternative actions available to an actor. In this thesis, institutional setting refers to the formal (voting or decision rules) and informal rules (iterated or repeated bargaining) of a bargaining game, which defines a particular negotiating situation in which n individuals (known as players) participate. This definition of the institutional setting is also in line with the traditional new institutionalist literature, which defines institutions as the formal and informal rules, such as procedures, codes of conduct, etc. (Hall and Taylor 1996, 938; March and Olsen 1989, 21).

Analyses focusing on the principals (member states) show only one part of the reality and totally ignore the agents (supranational institutions) and the European institutional arena. This is also in line with the work developed

1

by Garrett and Tsebelis (1996) who emphasize that in order to understand the European integration process, one has to analyze the effects of institutional rules among the European institutions. In this work the European integration process is defined as the gradual transfer of decision and legislation making from the national to the supranational (European) level, when the actors involved expect mutual advantages or gains from the policy transfer.[3] Since it is assumed that the intergovernmental and the supranational arenas are inextricably interrelated, the actors at the negotiating table are not only the national government representatives, but also officials of the Commission. This explains why the focus is on the Council-Commission axis in order to explain bargaining outcomes. Following the lead of Morrow (1994, 17) and Bueno de Mesquita (2000, 543), bargaining outcomes will be defined as results or consequences of the actors' policy decisions or choices, and it is assumed that the set of outcomes is exhaustive and mutually exclusive, that is, only one outcome takes place.

There are several ways of analyzing the European integration process. Some scientists have focused on the preferences of member states (Moravcsik 1993, 1997a, 1998) and their domestic constraints (König and Brauninger 2001; Hug and König 2002), others have considered the institutional structures (Garrett and Tsebelis 1996; Pierson 1996; Pollack 1997) and again others have emphasized the ideas and norms behind the integration process (Christiansen et al. 1999; Risse 2000). There is, however, a general consensus among scholars on European studies that governance in the EU consists essentially of negotiation and compromise among actors at the Council of the EU (Pfetsch 1998; Grande 2000). Accordingly the EU has been characterized as a "negotiated negotiation system" (Scharpf 1988), as a "multilateral inter-bureaucratic negotiation marathon" (Kohler-Koch 1996), and as a "negotiated democracy" (Lodge and Pfetsch 1998). The predominant mode for making decisions is bargaining at the Council, the "permanent negotiation forum" (Hayes-Renshaw and Wallace 1997) of the EU.[4] Consequently, several scholars stress the importance of the culture of compromise in the various stages of making decisions (Abélès and Bellier 1996; Hayes-Renshaw and Wallace 1997; Lewis 1998). How the EU is defined is of great importance, as the analysis of its decision-making system depends on the priorities the scientist places in defining the EU as a political system. The EU is considered to be a negotiating polity[5], since it deals with its problems by bargaining. Moreover, the system in its own constitutes a bargain, negotiated by the initial six member states in 1957. Since then, the manifold negotiations have modified the original structure and created a number of package deals on different policy fields and institutional arrangements. When there is governance by bargaining, national government representatives and officials of the Commission must reconcile their differing negotiating positions and arrive at a compromise solution encompassing the preferences of all the actors involved. This is often reached through consensus in last minute agreements.

In order to understand bargaining outcomes, the emphasis will be on the negotiation process. To avoid misunderstandings, it is important to define clearly what is meant by negotiation process. When analyzing the decision-making system of the EU one can look at structure or at process and how they interact. While structure focuses on the distribution of power resources among member states, process relates to bargaining behavior within a power structure. The game theory terminology can be used to illustrate how this works in practice: at the structural level, the important issue is how the cards and chips were distributed at the beginning of the game; whereas process level analyses concentrate on how the players play the hands they have been dealt (Keohane and Nye 1977, 21).

Negotiations in the EU are conceptualized as a process in which the actors agree on an agenda, explore initial policy positions, try to narrow gaps between actors' negotiating positions until a compromise solution is found, upon which all can agree. Bargaining implies a process that is completed when an agreement is reached. Negotiation and bargaining will be used as equivalent terms. A clear definition of the concept of negotiation[6] and its relevance to the concept of preferences is central to this dissertation in order to understand how actors bargain with each other. Bargaining is defined as a method used by actors to resolve or weaken their differences, when they disagree about which outcome represents the best solution for all the parties involved. In such situations agreement is most often reached by finding a compromise solution. Negotiations in the EU refer to intergovernmental (among national government representatives) and intra-institutional negotiations (between member states and the European Commission) at the Council framework, whether this be at working groups, Committee of Permanent Representatives (COREPER)[7] or ministerial level.

The purpose of this book is neither to duplicate overviews of the CFP nor to recapitulate narrative treatments of the European integration process. The aim is to comprehend how EU negotiations work theoretically and empirically and for this purpose a conceptual framework for analyzing internal negotiations in the EU will be provided and applied to a specific empirical field of research, namely the CFP. Two key negotiations in the fisheries policy field that led to the establishment of a common policy will be used to illustrate how the negotiation game in the EU is played in practice. In order to achieve this an integrative and a distributive bargaining situation were selected.[8] The term integrative bargaining will be used to describe a more cooperative bargaining situation, in which all involved parties wish to stretch the size of the pie bargained on. In contrast, distributive bargaining illustrates a more competitive and adverse process of dividing a pie. Actors involved in a negotiation have to try to distribute something, for example fish stocks, among themselves. In this second bargaining situation resources are fixed and limited and each side wants to maximize the share of the pie. Whereas the integrative bargaining situation focuses on a bargaining game, in which issues are linked in order to reach an agreement advantageous to all negotiating parties, the distributive bargaining situation illustrates a more adversarial process of dividing a resource allocation among themselves. In this case side-payments are necessary in order to compensate those member states that considered the agreement to be disadvantageous to themselves. Thus, the analysis of these two key negotiations reveals much about methods of making decisions in the EU and about the way bargaining outcomes are reached. The CFP will be used as a means to an end, that is to say, to illustrate which variables play a role in explaining bargaining outcomes in EU negotiations. In sum, the analysis of both bargaining situations attempts to demonstrate how the actors' preferences and the institutional setting interact with each other, how integrative and distributive bargaining games may differ from each other, and how issues can be linked in the EU negotiating polity.

The analytical framework used is based on the rational-choice tradition. Negotiations in the EU are a situation where n actors (or players) try to agree on one of m possible solutions. The negotiation process has a multitude of actors (national government representatives and the European Commission) each with his/her preferences. Referring to an actor having a certain preference means that an actor prefers one alternative to another and in consequence tries to maximize his/her utility, when s/he tries to achieve outcomes that are as close as possible to his/her preferred negotiating position. Actors will be modeled as individuals with bounded rationality.[9] Rational means here that each player chooses the alternative that gives him/her the highest utility (greatest benefit). Every actor considers the consequences of all the possible options from which a choice can be made, and ranks these outcomes in order of preference, from the most preferred down to the least wished. This preference ordering has to be consistent in the sense that if A is preferred to B and B is preferred to C, then A must also be preferred to C. Perfect rationality, however, is a fiction. There is no such thing as a perfectly straight line. Hence political actors try to push through their preferences, but they act under bounded rationality, that is they are not completely informed and do not dispose over the cognitive capacities to maximize everywhere their preferences. National government representatives, when attempting to maximize their preferences, are constrained by their time horizons and by the institutional setting. The Commission, when trying to maximize its preferences, is constrained by the preferences of member states.

This study intends to analyze, theoretically and empirically, why bargaining outcomes in the EU are what they are. For this purpose, a conceptual framework will be constructed. Thereby its major function will be to identify a set of variables that presumably explain the bargaining outcomes. Thus, the research design is quite simple. A set of independent variables (preferences of national government representatives, preferences of the Commission, and the institutional setting) is expected to determine the bargaining outcomes (the dependent variable). The key research question is when the preferences of the Commission and the institutional setting are assumed to be independent variables, how and to what extent do they affect the bargaining outcomes.

METHODOLOGICAL PROBLEMS WHEN ANALYZING NEGOTIATIONS IN THE EU

Negotiations in the EU are an intergovernmental and inter-institutional phenomenon that takes place *in camera* leaving the scholar outside the negotiation room. Consequently, any study of negotiations in the EU turns into a reconstruction of what takes places inside a closed room (Friis 1997, 19). There are several methodological problems interconnected with the analysis of negotiations in the EU. They can be summarized as follows: the closed door problem, the strategy-problem and the sensitivity problem.

Any scientist analyzing EU negotiations is subjected to the closed door problem. Public administration is generally a closed field for the external researcher and the administration of the EU is no exception to bureaucratic secretiveness. Until the transparency decision of 1992 at the Edinburgh Council, the Council minutes were subject to a thirty-year law of restriction. Now the EU must ensure access to documents of the European Parliament, Council of the EU and European Commission. Since that decision was taken, improving the right of access to internal documents is considered to be an integral part of transparency and an attempt to lift the veil of secrecy surrounding the Council (Sherrington 2000, 69). The applicant, however, needs a great deal of patience and valid reasons before s/he attains the desired access.

Nevertheless, the lack of transparency is still considered part and parcel of the European integration process. In the European studies literature it is frequently argued that interstate bargains are only possible because decision-makers operate in closed session. Thus secrecy is inherent in EU negotiations and even seen as indispensable in many bargaining circumstances. Hayes-Renshaw and Wallace (1997, 7) point out that the secrecy of the Council's negotiations and decisions allows the ministers to speak in unvarnished terms and also to use arguments that would not be so easy to repeat if the negotiations were public. The consequence is quite clear: one can seldom rely on transcripts from the negotiation room. A possible solution could be to base analyses on the various statements by political actors (prime ministers, agriculture ministers, and so on) which characterize the period before, during and after the negotiation process. The value of these sources, however, is highly questionable. Indeed it could be argued that they are part of the EU negotiation game, that is part of the bargaining strategies employed by actors to influence the game. In the newsroom the ministers not only try to exert pressure on the other member states, but also try to change the overall mood in their home country. Sentences like "we really fought hard on this issue, but the others simply would not accept our view" are often used to explain a certain outcome to the domestic constituencies. It is clear that two games are

being played: one for the national public opinion and another in the negotiation room (see Spence 1995, 376). Thus a double-edged diplomacy is part
of the game.

Finally, negotiations in the EU are generally seen as a sensitive research
area. Friis (1997, 21–22) asserts that scholars exposing the EU are more or
less denouncing a crucial aspect of it: the horse-trading or the Bazaar tactic.[10]
Research on negotiations in the EU emerges inherently as a threat, since
analyzing the negotiation process is like opening Pandora's box. Furthermore, the EU is an ongoing negotiation game, where any post-negotiation
can develop into pre-negotiation for the next bargaining game. This explains
why a high reluctance still exists in guaranteeing access to those internal
Council documents, in which policy or bargaining positions are clearly
dissected.

Some authors (Wallace and Hayes-Renshaw 2001; Friis 1997) assume
that scientific interest in analyzing negotiations at the Council is insignifi-
cant and explain this with the fact that this institution has been seen
primarily as the defender of the interests of the member states abroad rather
than that of the ideals of European integration. There are, however, many
studies now focusing on the Council and on negotiations in the EU (Bueno
de Mesquita and Stokman 1994; Bulmer and Wessels 1987; Hayes-
Renshaw and Wallace 1997; Sherrington 2000; Westlake 1995). The
problem with many of these studies is that their mode of analysis consists
of a detailed description of the structures and procedures of the Council.
Bueno de Mesquita and Stokman's (1994) study is an exception, where,
assuming the pivotal role of the Council, they tested a number of policy-
making models. Finally, a great part of the literature on EU negotiations is
actually written by insiders sharing their knowledge (Farnell and Elles 1984;
Holden 1996).

To sum up, the inherent methodological problems of analyzing the EU
negotiation process appears to represent the state of the art: the problems are
largely side-stepped by staying at the descriptive level and leaving the more
process-oriented analysis to those directly involved in the negotiations.

GAME THEORY: A USEFUL TOOL TO BYPASS THE METHODOLOGICAL PROBLEMS

It is well known that European studies lack investigations based on "hard"
primary sources (Moravcsik 1997b). The present work is an attempt to close
this gap. The methodology of this study consists of a combination of the use
of primary sources, which include Council minutes and internal working
groups documents, and secondary sources. Primary sources will be used in
order to complete, corroborate and in some cases to refute the secondary
sources. By doing so parts of the strategy-problem can be side-stepped and
also the negotiation process can be reconstructed. Data collected from
October to December 1999, during a training session at the Council, are
used to try to reconstruct the negotiating positions of the actors involved.

The Fisheries Council negotiations from 1970 until 1983 have been described by two Commission practitioners (Farnell and Elles 1984) and also in a scientific study (Leigh 1983). Both studies, however, do not go beyond the descriptive level.

It is assumed that part of the secretiveness of the Council's negotiations can be broken down through direct access to internal negotiation documents. This, however, shows only one part of the reality, actors' negotiating positions and bargaining outcomes. The reality of EU negotiations is more complex than it may appear at first sight. Ministerial meetings are only the peak of much larger and broader negotiation processes. Negotiations in the EU are not based on a single negotiation, but are mainly a process, a step-by-step game which starts at the Working Group level, and goes on to the COREPER.

In EU negotiations a distinction is made between "A and B points". "A points" ("Agreed points") are sent to the ministers *en bloc* and are the first item on the agenda passed without discussion. Only "B points" (non-agreed points), where agreement could not be reached at the lower level, are at stake, and the decision is delegated to the higher (minister) level. It follows that many of the points for discussion in the agenda have already been decided before the sectoral Council meetings take place. In this case, the Council's presidency simply states that "A-points" have been adopted. Whereas in the case of "B-points" the outstanding points are repeated and the bargaining positions of the different member states stated (cf. Westlake 1995, 87). At this point of negotiations, discussion and amendment of the original proposal usually continues until a compromise formula is found which benefits all the actors involved or at least with which all can live.

In bargaining games, actors' preferences diverge, otherwise it would not be necessary to negotiate. The game theory may be a useful methodology for analyzing bargaining games, since it is the only existing analytical framework for explaining interdependent decisions among rational actors.[11] Interdependent decisions, in turn, take place where an individual's payoff depends on the action chosen by the other player and when the decisions of two or more players jointly determine the outcome of a situation. Game theoretical approaches assume that players involved in an interaction will be able to choose from a set of strategies, and that the choice will be made to maximize individual payoffs. The optimal solution to a game is the strategy combination that gives a payoff to all players and which can not be improved upon by any other alternative without making any player worse off. When explaining the negotiation process in the EU, bargaining strategies, defined as an actor's plan of action or strategy in a given bargaining game, also have to be taken into consideration, since they may determine outcomes. Negotiations in the EU are a good empirical field of observation for the application of game theoretical concepts, since the European integration process can be described as a cooperative game. Member states coordinate their strategies through binding agreements before and during the negotiations, and also act jointly to maximize their gains. The main issue in a cooperative game consists in reaching an agreement that is Pareto-superior to the status quo, that is those agreements which make everyone better and no one worse off.

The advantage of using the game theory as a methodology is that it forces one to specify the structure of the game: who are the players at the negotiating table, what preferences do they have, how strongly each player prefers one outcome to another when comparing two or three possible outcomes of the negotiation and how does the institutional setting constrain their behavior and influence bargaining outcomes. Furthermore, one has to specify how these preferences lead to results and which variables explain the bargaining outcomes. Game theory also allows one to analyze the different bargaining moves—that is how the bargaining game progresses from one stage to another, beginning with the initial stage of the game through to the final move—in a more formalized way. At the present, process and actor oriented studies are still lacking in the field of European studies. By focusing on who the players are, which preferences they have and on how and to what extent the structure of the game influences the bargaining behavior of member states, an attempt will be made to explain how bargaining outcomes are reached among the actors involved in a negotiation process. For this purpose one needs game theoretic concepts, since they provide the scientist with a methodology that allows him/her to formalize social structures and simultaneously to examine the effects of structure on individual decision.

THE ROLE OF THEORY IN POLITICAL SCIENCE

In political science, explaining bargaining outcomes or other empirical evidence depends on two components: facts and theoretical perspective. Some scientists assume that knowing the facts allows one to adequately explain what will happen in the future. But focusing merely on a catalogue of facts may lead to a distorted view of reality, as the facts may not tell "why" they evolved as they did. All observations of reality are "theory-impregnated" (Popper 1994). Not only the selection of facts is shaped by the circulating theories, but also the way to interpret those facts is influenced by certain theoretical perspectives (cf. Bueno de Mesquita 2000, 135; Zürn 1994, 94).

Examining political phenomena from different conceptual and theoretical approaches is a way of reaching alternative explanations. The theoretical framework of this thesis is based on the premise that science should try to find an answer to the question "why" political outcomes occurred. For this purpose theories, stating how the variables may influence each other, are necessary. Broadly speaking, a theory is an abstraction or simplification of a complex reality, a *"Gedankensystem"* (Feyerabend 1981). Each science needs theories to analyze facts. Using a theory to explain EU negotiations means employing a useful set of tools to organize and interpret empirical data.

The basis of any theory must be a logical and consistent framework based on premises or assumptions that do not contradict each other. Logical coherence also implies precise definition of the key concepts used. Furthermore, in the Popperian sense, knowledge is always preliminary, that is theoretical propositions should be framed in such a way that they are falsifiable. According to the criterion of falsifiability, also called the criterion of testability of a theoretical proposition, a theory is made plausible by rigorous empirical testing. Testing a theory is to try to fault it, because if a theory cannot be falsified *a priori*, it is not testable (Popper 1994, 89). Only empirically refutable propositions should be regarded as scientific. In short, Popper's (1994, 159) view of a scientific method consists of four steps: first, a problem is selected; second, a scientist should try to solve that problem by proposing a theory seen as a tentative solution; third, a critical discussion of the theory is required to eliminate its errors, because knowledge can only grow through error-elimination; fourth, this critical discussion, in turn, leads to new problems. These four steps can be summed up as: problems, theories, criticisms, new problems.

Popper's rule that one had to specify a potential falsifier to the hard core of a theory has been strongly criticized, mainly by Lakatos (1970, 1999), who counter-argues that the greatest victories in science were verifications and not falsifications. The work of Lakatos draws our attention to a weak point in Popper's scientific method, namely the fact that theories are not simply discarded, but redefined and reformulated. On the one hand, theories, after being admitted, must be built gradually, progressing from less formal to more formal stages of development. On the other hand, scientists do not hold on to cherished theories unconditionally: some theories fail and consequently are substituted by others (Williamson 2000, 607). Science proceeds by a deliberate process of developing, testing, refining and sometimes overturning hypotheses and theories that explain the patterns of cause and effect, which have become apparent. A new theory must be able to explain why its predecessor was not successful. One needs a series of theories, in which each one is reached by adding some auxiliary clauses in order to accommodate certain anomalies and produce new predictions.

The central issue is whether there are some standard rules to be observed. Solow (2001, 111–112) has set up a catalogue of criteria for practicing good economics, namely keep it simple, get it right, and make it plausible. The first criterion refers to the fact that reality is much too complex to be grasped in total. Second, getting it right means essentially to translate concepts into accurate diagrams and to make sure that further logical operations are correctly executed and verified. Using diagrams or equations allows one to be precise about the assumptions and logical consistency of a model or of a conceptual framework. One can also go one step further and build sophisticated mathematical models, but this is not a must. Finally, a conceptual framework has to maintain a plausible contact with the analyzed phenomena. It can be mechanically correct, but must also be suited to the analyzed subject.

In order to apply these criteria to the conceptual framework for analyzing negotiations in the EU they need to be adapted into political concepts. By definition a model is a representation of reality focusing on key intervening variables. One of the advantages of modeling a negotiation process is that it allows one to reduce the number of elements for capturing decision-making in a certain polity. The modeler tries to mirror the reality by simplifying it. Every analysis of a process or an outcome is partial, that is it points out some variables of a situation and excludes others. Furthermore, since theory and empirical evidence are intrinsically correlated, conceptual frameworks must be subjected to rigorous empirical testing, otherwise they remain merely amusing mathematical or theoretical exercises. The gulf between theoretical and empirical work can only be bridged by putting all these ideas into practice, and this has not always been done in the past.

The overall methodology of this thesis is guided by the attempt to offer theoretical and empirical insights into the bargaining process of the EU negotiating polity. Its primary aim is to be explanatory. It tries to present schematically the independent variables that may explain bargaining outcomes, since in the long run, a conceptualization of the negotiation process in the EU with its negotiation logic and strategic moves, may be much more useful than just describing all the negotiation details.

SEARCHING FOR THEORETICAL CONCEPTUALIZATIONS OF THE EU BARGAINING GAME

There are different theoretical conceptualizations of the bargaining game in the EU. One can use the classical integration theories, from state-centric to new institutionalist approaches, or apply game theoretic concepts for explaining the strategic behavior of actors at the negotiating table. Before turning to the theoretical approach used in this work, it may be useful to give a short overview of the different theoretical conceptualizations of the EU bargaining game.

Classical integration theories neglected the negotiation process, because their central goal was not to explain the outcome of, for instance, a negotiation on agricultural prices, but to analyze the overall integration process. Neofunctionalism assumed that the European integration process would occur automatically and gain weight gradually, like a snowball or a spill-over process (Haas 1961). The negotiation process in the EU was seen as a supranational style of bargaining, in which supranational institutions and interest groups played a central role, while the member states were relegated to the back seat. In turn, intergovernmentalism also remained at the process level by arguing when and why the European integration process stagnated or progressed, but without analyzing the bargaining process. The intergovernmentalists merely exchanged the neofunctionalists' spill-over-process for the inter-state bargaining process (Hoffman 1966, 1982). The integration process was made dependent upon the preference convergence between the member states; hence explicitly excluding the role played by the European Commission. Both theoretical approaches argued that bargaining was irrelevant, because the result would always be the lowest common denominator.

Lindberg and Scheingold (1970) were the first to take a closer look at the negotiation process and tried to draw up a strategy-typology of the various bargaining moves (log-rolling, side-payments, package deals and issue linkage). This seminal work also had two key problems. First, the analytical approach was still strongly biased towards the European Commission. The strategy-typology showed the various strategies used by the Commission to speed up the integration process, but there were no analytical criteria for how to analyze the behavior of the member states. The central question of how the preference formation of member states actually takes place was simply ignored and the analysis of the bargaining choices of the member states neglected. Second, Lindberg and Scheingold looked upon the European Commission as a neutral actor concentrating exclusively on advancing the common interest. Nevertheless, this assumption is hardly credible, as this work will try to demonstrate. They simply oversaw that above all the Commission is a rational maximizer bureaucracy.

After a so-called period of "dark-ages" in the 1980s, the appearance of Moravcsik's liberal intergovernmentalism in the 1990s gave the theoretical debate a new momentum, with almost everyone defining himself/herself in opposition to this theoretical approach. Moravcsik (1991, 1993, 1998, 1999 [together with Nicolaidis]) tried to explain the outcomes of the EU's grand institutional bargains (the Treaty of Rome, the Single European Act, the Treaty of the European Union, the European Monetary System and the Amsterdam Treaty) as a three-stage process. First, at the domestic level national governments define a set of interests for policy cooperation at the EU level (preference formation). Second, they bargain among themselves to defend those interests (interstate bargaining). Third, institutional delegation reflects the desire of the member states for credible commitments. The bargaining process is seen as a result of the interaction between national preferences and bargaining power.

While for state-centric approaches the preferences of member states are the central variable to explain bargaining outcomes, institutional approaches towards European integration focus essentially on two research questions, how institutions originate and change, and how the relationship between institutions and the individual behavior of political actors interacts (Hall and Taylor 1996). Bargaining outcomes are interpreted as the composite effect of a large variety of institutional variables. In the last few years, debates have concentrated on the three different strands of new institutionalism, namely rational choice, historical and sociological institutionalism. They all assume that institutions affect outcomes, that is the institutional set of a negotiation process is more than a passive arena, being characterized by independent logics (Aspinwall and Schneider 2000; Pollack 1997; Garrett and Tsebelis 1996). Once states set up institutions they also run the risk of creating norms, rules, ideas, and dead-weight, which all act as intervening variables between preferences, power and outcome (Bulmer 1994, 1998; Pierson 1996).

Each of the three different strands, however, has a distinct definition of institutions and how they "matter" in the study of politics. Oversimplifying slightly: for rational choice institutionalists, institutions constitute an intervening variable, which can affect an actor's preference, but not determine it; for historical institutionalists institutions shape the preferences of actors, and finally sociological institutionalists assert that institutions depend on larger "macro level" variables like society and culture (cf. Koelble 1995, 232).[12] All these approaches have the problem that they do not show clearly how structure really matters and how those structures are indeed able to affect the behavior of actors.

Recently two reviews, *International Negotiation* (1998, 2002) and *Journal of European Public Policy* (2000, 2002) dedicated special issues to the analysis of negotiations in the EU. The articles collected in the latter deal with three different aspects of negotiation: the EU as a negotiation process, the EU as a system, and the EU as a negotiated order. Under the EU as a negotiation system, Elgström and Smith (2000, 675) identify three fundamental characteristics: interdependence of actors, regularities of interactions, and the existence of (informal and formal) rules or institutions. They do not, however, really give a definition, but rather an identification of the key features and remain, as in many other studies, at the level of the affirmation of the uniqueness of the EU negotiation system. Finally, two approaches can be distinguished in the EU as a negotiated order: the structural approach describes the EU as a reflection of a set of forces in the European arena through which a convergence of expectations and resources around the EU's negotiation system has taken place. The second approach concentrates on the processes, that is, on the way of settlement, maintenance and modification of a negotiated order.

In the literature of last decade there has also been a tendency towards the application of the game theory to the EU negotiation process. First attempts were made by Putnam (1988), Tsebelis (1990), Bueno de Mesquita and Stokman (1994), Schneider and Cederman (1994), and Scharpf (1997). Putnam (1988) was the first to conceive international negotiations as a two-level game, meaning that in practice actors have to bargain in two arenas simultaneously, at the national and international level: on the one hand, a national government representative negotiates with other foreign countries (Level I), and, on the other, s/he is in a bargaining situation with relevant domestic constituencies (Level II). As Hayes-Renshaw and Wallace (1997, 23) state, this provides opportunities for political manipulations between the two levels; under the cover of international pressures, national government representatives can sometimes take a more innovative and risky position than when handling at the national level. A collection of essays edited by Evans *et al.* (1993) develops, enhances and expands Putnam's metaphor on international negotiation analysis through the more detailed exploration of issues other than the economics of the interaction between the two levels. More recently Patterson (1997) analyzed the conditions under which a shift occurred in the common agricultural policy using a "three-level game" approach, which involved parallel negotiations within the General Agreement on Tariffs and Trade (GATT) with the EU as one party, among the EU member states, and between various interests within key member states. Tsebelis (1990) offered an analysis of what he calls "nested games", meaning that an actor's choice may appear sub-optimal or "irrational" at the European level, because the observer's perspective is incomplete. In order to understand the actor's negotiating position, one has to focus not only on one specific game but on a whole network of games, because what seems to be sub-optimal from the perspective of analyzing only one game, is indeed optimal when the whole network of games is taken into account. The analysis of Bueno de Mesquita and Stokman (1994) focuses on the effects of the influence of the strategies chosen by actors involved in the decision-making process of the EU and on the outcomes of this process. They both analyzed two directly comparable models: the non-cooperative model of Bueno de Mesquita expected utility and the cooperative exchange model of voting positions of Stokman and Van Oosten. Both models are based on the same three variables: the capability of actors to determine the outcomes of decisions, the importance of the issues for the actors and the actors' preferred outcomes (policy positions of the actors). They modeled, however, the decision-making process by focusing on the issue how member states bargain with each other at the Council of the EU, and ignoring the other EU institutions. Schneider and Cederman (1994), focusing on the strategic use of uncertainty in international negotiations, showed how the threat of one conservative obstructionist member state in the EU negotiations can lead to a breakdown of the negotiations. They argued that the decision-making structure of the EU enables a laggard state to blackmail all the other members. Scharpf (1997) created an actor-centered institutionalism, in which policy outcomes are seen as the result of interactions of bounded rational actors, whose preferences and bargaining power are largely, but not totally, shaped by the institutionalized norms.

Although so many different approaches are already available, the logical question is whether a new approach is required for conceptualizing negotiations in the EU.

USING A MULTI-THEORETICAL APPROACH IN THE ANALYSIS OF EU-NEGOTIATIONS

Theoretical and conceptual approaches to the study of the European integration process vary and they can be divided into categories on the basis of what they are seeking to analyze and explain (Cram et al. 1999, 8). In this respect, Nugent (1999, 10) distinguishes between: grand theory, which tries to explain the integration process as a whole, middle range or meso theory, which attempts to explain aspects of the functioning of the EU, and conceptualizations whose aim is to capture the essence of the EU in conceptual terms. The present book tries to explain one aspect of the functioning of the EU, namely, how bargaining outcomes are reached in the Council of the EU. It can be considered as a middle range theoretical approach.

In order to be able to analyze how bargaining outcomes take place at the EU negotiating polity, it is necessary to use a theoretical approach, which to a certain extent can be found in the literature, but which has still to be put together. Some scientists on European studies state that the complexity of governance in the EU can no longer be explained by a single theory or model. European studies scientists have to learn to link conceptual frameworks, theories and multiple models in order to be able to combine them into a coherent approach (Richardson 1996; Moravcsik 1998). Furthermore, as has been emphasized, few scientists employ a multi-theoretical approach, and the different theoretical approaches tend to ignore or to play down the insights offered by the others (Pollack 1997, 430; Moravcsik 1998, 17).

This work attempts to link liberal intergovernmentalism with new institutionalism and with public choice. Putting these three approaches together may help to understand the relationship between member states' preferences, which are central for liberal intergovernmentalism, and the role played by the formal and informal rules in bargaining games, which are a crucial issue in new institutionalist and public choice approaches. Moravcsik's liberal intergovernmentalism can be seen as a useful springboard for analyzing negotiations in the EU at the European Council, because it opens the black box of state interest formation and conceptualizes clearly the bargaining power of the member states. Although liberal intergovernmentalism is one of the few theories which has tried to grapple with the negotiation process in the EU, Moravcsik attempted to create an elegant theory with very few variables. As a result, he ends up with a too simplified black and white picture. This explains why liberal intergovernmentalism has some limitations when analyzing bargaining processes at the Council of the EU. A more important shortcoming of liberal intergovernmentalism, however, is that it maintains that supranational institutions (such as the European Commission) have had only a marginal effect, not only on the negotiation process, but also on agenda-setting, legislation, implementation and enforcement of Community policies.

In the conceptual framework developed for analyzing negotiations in the EU, not only the preferences of member states, but also the preferences of the Commission will be dissected. Besides its role as an agent of member states, the Commission started to develop its own preferences and tries to conceal them through its formal agenda-setting power. Hence one has to go back to the principal-agent framework, which offers useful clues to analyze how a relationship will face problems of accountability when the preferences of the principals and agents do not coincide. Furthermore, it is asserted that the institutional setting may matter. New institutionalist and public choice approaches may be useful in order to be able to answer the central question of this thesis whether, and to what extent the preferences of the Commission and the institutional setting affect bargaining outcomes in the EU.

In other words, in contrast to state-centric approaches, it is assumed that the negotiation process in the EU has more than the preferences of member states and bargaining power as intervening variables of the bargaining outcomes. Therefore, the conceptual framework for analyzing the negotiation process will be based on the following variables: a clear identification of the players (member states and European Commission), the configuration of actors' preferences, and the institutional setting. On the basis of liberal intergovernmentalism, new institutionalism and public choice not only the behavior of actors, but also the effects of the institutional setting will become more transparent and an explanation for bargaining outcomes will be found. Such a conceptual framework for analyzing internal negotiations may be a new way to analyze integrative and distributive bargaining situations in the EU. Moreover, it may enable the empirical testing of European integration, because, as is frequently pointed out in literature, the empirical testing has not kept up with theoretical advances (Peterson 2001).

THE ROAD TO NEW FINDINGS

After the attempt described above to clarify the aim and methodology of this dissertation, its concrete structure will be now delineated. The book is divided into four major parts. Chapter one takes a more empirical approach focusing on the settlement of the CFP. Nowadays the competence of fisheries policy has been transferred from the national to the supranational level. A central issue of this chapter is to explain how fisheries came into the agenda of the EU. A short overview of the four different pillars of the CFP will be given, followed by the external and internal policy input variables that explain the policy transfer to the EU. In order to understand why collective action was necessary this chapter deals shortly with the open access character of fisheries until the beginning of the 80s and furthermore it describes the tragedy of the commons in the fisheries sector.

Chapter two presents the central concepts of the conceptual framework for analyzing EU negotiations. After some introductory remarks on the status of the framework, the variables playing a role within it will be introduced. After defining who the actors are, the next step defines their preferences and tries to specify what influences the preference formation of national government representatives and of the European Commission. In line with rational-choice studies, preferences of member states were taken as given exogenously. Furthermore, the black box of preference formation will be opened using the following four indicators: time horizons of decision-makers, interest groups and their organizational effectiveness, level of politicization of one issue, and national parliaments as potential internal veto players.

After having defined the preferences of member states, the next step will consist in trying to define the preferences of the European Commission, since one of the main problems of the European studies literature is the failure to identify and explain the preferences of supranational actors. It is assumed that the preference formation of the European Commission is influenced by its being a representative of the European Community (EC) and also a rational maximizer bureaucracy. The institutional setting is the next variable used to explain bargaining outcomes. It will be distinguished between formal and informal rules. Formal rules are constituted by decision or voting rules. The formal rule unanimity voting is used as the constant parameter for analyzing two negotiating situations. Informal rules refer to the variable iterated bargaining, that is the influence of the time dimension in bargaining outcomes. This chapter closes with a differentiation of the integrative and distributive bargaining games and an estimation of the preferences of member states in a policy space.

After establishing who is playing the game, the players' preferences, what they are playing for and what the rules of the game are, the two most interesting negotiations in the fisheries policy field were selected to find out how actors' preferences and the institutional setting interact empirically when unanimity is the voting procedure, and there is a variation on the number of actors involved. Chapter three deals with an integrative bargaining game: all actors expected to benefit from agreement in the negotiations on the settlement of the structural policy and of the common market organization for fish products. Chapter four discusses a more controversial bargaining process, a distributive bargaining game. Here the focus is on bargaining over the division of potential benefits with the analysis of the acrimonious negotiations on the conservation and management policy. The distributive bargaining situation is more controversial by nature, because each member state's concern was to maximize its own benefits and minimize its own losses. In both chapters, after a brief overview of the actors' negotiating positions at the opening bid, the main focus is on the bargaining process in order to find out whether different negotiating tools are used to overcome the unanimity trap, and how the independent variables used in the conceptual framework influence the bargaining outcomes. While in the integrative bargaining game the focus is on coalition-building, in the distributive bargaining game veto players dominate the different bargaining rounds. Consequently, the sequence of moves will be modeled as an extensive-form game.

The conclusion will draw together all the arguments, the diverse strands of the analysis will be assembled and present the argument about which variables play a role in the EU bargaining process. In short, this dissertation deals with a vast process. It involves a large amount of piecing together, analyzing and explaining events on the basis of primary and secondary documentary sources. It is an attempt to provide an insight into the complex and multifaceted negotiation process in the EU.

CHAPTER 1

The Settlement of the Common Fisheries Policy

Trying to Cope with a Changing Environment

SOME BASIC FACTS: THE FOUR PILLARS OF THE COMMON FISHERIES POLICY

The EU engages in four main activities, namely establishment of a common market, adoption of common policies, creation of a common foreign and security policy and study of its own constitutional development. Since decisions in the areas of political cooperation and constitutional development require the participation of high ranking political leaders and are subject to political debates within each member state, these issues are in general better known than the other two fields. Although the categories related to the common market and the areas of common policies constitute the main part of EU decisions, little is known about their settlement and which variables play a role during the negotiation process (Pierce 1994, 8).

The present chapter gives a general overview of the four different pillars and policy instruments of the common fisheries policy (CFP), tries to explain why the policy transfer, defined as the transfer of decision and legislation making from the national to the European level, took place at a certain point in time, why this policy structure came into existence and what is the state of the art in the scientific literature on the CFP. A common policy in the EU is the sum of the European Community's legislation (directives, regulations, decisions, and recommendations) put into force. Any single policy has a certain number of pillars and instruments, which together constitute its centerpiece. The CFP is divided into four distinct pillars: structural policy, common organization of the market, fisheries agreements with third countries and resource conservation and management system. Each of these pillars was developed in response to different external and internal outputs arising at different periods. *Figure 1* provides a general overview of these four different pillars of the CFP. Thereby, the numbers in brackets denote the year in which the individual policy segment was settled.

The first pillars of the CFP, the structural policy and the common organization of the market, were introduced in 1970. The structural policy aims

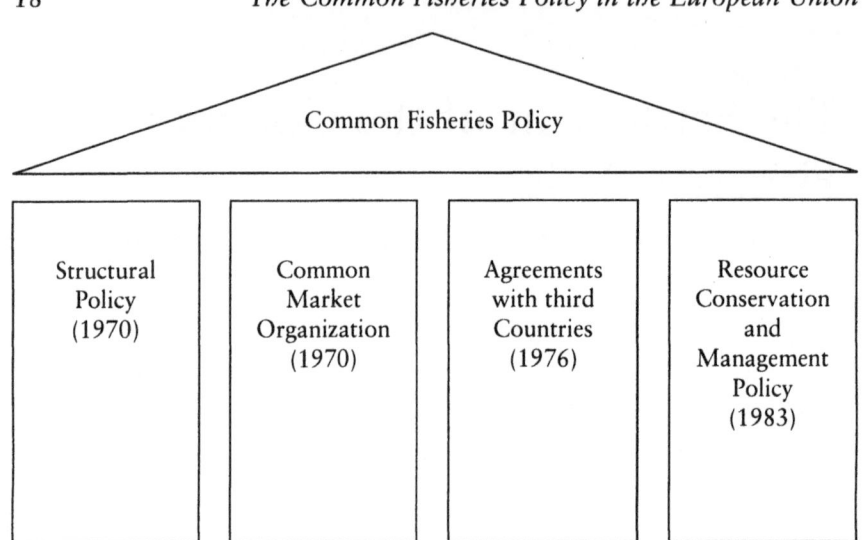

Fig re 1. The four pillars of the Common Fisheries Policy

to eliminate excess fishing capacity by giving financial assistance to restructuring and modernizing the fishing fleet of EU member states. With the adoption of a regulation outlining a common structural policy, member states were required to coordinate their national structural policies. They were allowed to grant financial aid to their national fishing industries insofar as fishing vessel construction and modernization were concerned. Financial assistance was made available directly from the European Agricultural Guidance and Guarantee Fund (EAGGF). In 1983 a new package of structural measures was adopted consisting of a system of further financial aid for the removal of vessels from national fishing fleets and a new scheme for the provision of vessel construction and modernization subsidies. In this way, a new policy instrument, the multi-annual guidance programs (MAGPs) was settled. However, these EC's structural programmes have been often criticized, because they have failed to adequately control real growth in the catching capacity of the EC's fishing fleet. Furthermore, the guidelines on state aids for fishing vessel construction has not been sufficiently tight so that national state aids exceeding specific rates were approved (cf. Hatcher 2000, 130).

The second pillar of the CFP, the common organization of the market, established a common market for fish products. Basically, it takes over the principles applied to the Common Agricultural Policy (CAP) of the EC. Accordingly its main aim is to stabilize the fish markets, to secure supplies and to ensure that consumer prices are not too high. The CAP was justified as making Europe self-subsistent and at preventing people from moving from the country. Therefore, the EU increased assistance to farmers to increase production and to give them better incomes. The same policy measures were applied to the CFP leading to a rapid expansion of the European fishing fleet. By and large, the common organization of the market has four components: common marketing standards for fresh products; a price support system which sets minimum prices below which fish products cannot be sold; compensation payments for products withdrawn from the market at minimum price levels, this is made by producers' organizations, which are voluntary associations of fishermen set up to help stabilize markets; and rules for trade with non-EC countries.

The external policy, the third pillar of the CFP, was settled in 1976. It consists primarily of agreements with third countries negotiated at the supra-national level between the European Commission, which has a mandate from the member states, and the third countries. Since then a variety of EC fisheries agreements with third countries have been created. Three different forms of Community fisheries agreements can be distinguished:

(a) *agreements with financial compensation*, in which the EU acquires fishing rights in exchange for overall financial compensation paid from the Community budget and the ship owners, this type of agreement is in operation between the EU and the African and Caribbean Countries;

(b) *reciprocal agreements*, in which the EU grants fishing rights to third countries in member states' fishing zones in exchange for similar rights for Community vessels in those countries waters, for example the agreements with Norway, the Baltic countries, the Faeroe Islands, and Iceland;

(c) *"second generation" agreements* which are based on incentives for setting up joint ventures which can develop their fishing activities in the waters of third countries, at the present there is only one agreement of this type, with Argentina (IFREMER 1999, 4–5; Holden 1996, 35).

The last pillar of the CFP the conservation and management of fish stocks aims to limit fishing activities by setting annually maximum catches and simultaneously decreasing the number of fishing vessels in Community waters. This pillar is based on the concepts of total allowable catches (TACs)[1], which specify annually the maximum quantity of each fish stock that can be fished, and a national quota system, which distributes the volume of such catches among the member states. TACs and national quota systems can be considered the centrepiece of the conservation measures, which also include permitted gear[2], restrictions on the allowed time period for fishing, and the fishing areas. The distribution of fishery resources is built on the principle of relative stability, which in turn is based on three key factors. These are historic catches, special allowances based on "vital need" or "Hague preferences"[3], and compensation for jurisdictional losses after the introduction of the exclusive economic zones (Coffey 1995, 15; Karagiannakos 1996, 236).

There is widespread agreement in the literature that the policy transfer to the supranational level could not solve the problem of resource scarcity (Holden 1996; Karagiannakos 1995; Payne 2000; Symes 1997). Analyzing the data of the North Sea demersal fisheries, Karagiannakos (1996) demonstrated that the TAC system does not contribute significantly to maintain fish stocks and conserve the resources. Even the European Commission (2001) recognizes in its *Green Paper on The future of the common fisheries policy* that the conservation and management policy has not achieved a sustainable exploitation of fisheries stocks.

EXPLAINING THE POLICY TRANSFER TO THE EU: EXTERNAL AND INTERNAL POLICY INPUT VARIABLES

One of the fundamental problems when analyzing common policies is that studies providing a good overview of how, and why a common policy has come into existence, can rarely be found. Some scientists came to the conclusion that it is almost impossible to trace the origin of a proposal (Richardson 1996; Symes 1997). In the case of the CFP, it has been argued that the legal origins of it were "obscure and unconvincing" (Symes 1997).

This chapter is an attempt to dissect the labyrinth of policy initiation and tries to explain why particular events took place when they did. *Table 1* summarizes the most important decisions in the settlement of the CFP. The origins of the CFP go back to 1970 with the settlement of the structural policy and the common market organization and the introduction of the principle of equal access to the territorial waters of member states. With the first enlargement of the EC, the act of accession for Great Britain and Ireland introduced a ten years restriction on this principal of equal access. The next milestone was the agreement at the Hague in 1976 on the creation of the 200 mile Exclusive Economic Zone (EEZ) which grants sovereign rights of access, exploitation, and management within that zone for the EC member states and embraces about 90 percent of the resources. Furthermore, the legal exclusive competence for international fisheries agreements or relations with third countries was also transferred to the EC. External and internal input variables provided the momentum behind both decisions.

At this time, the first attempt to introduce a management system to determine the allowable catches failed. This was just the beginning of protracted negotiations that started in 1976 and lasted until 1983, when an agreement on the conservation and management policy was finally found. During the negotiations, the accession of Greece had little impact on the CFP altogether, because the Mediterranean Sea is not a part of the agreed common policy. Only the EC's structural policy is important for Greece, as it needed to modernize its fleet and shore based infrastructure.

The adoption of the United Nations Convention on the Law of the Sea (UNCLOS) in 1982 and the accession negotiations with Portugal and Spain led finally to the establishment of the last pillar of the CFP: the conservation and the management policy for fish resources. Since the accession of two

Table 1. Key Dates in the Development of the Common Fisheries Policy: 1970–2002.

Date	EU	Development	Regulation No.
1970	EC 6	• Establishment of the two first pillars of the CFP, namely the structural policy and the common market organization • The principle of *"equal access"* to the territorial waters of member states is introduced	2141/70 2142/70
1972	EC 6	• The Act of Accession for Great Britain and Ireland introduced ten years restrictions to the principle of "equal access" to fisheries resources	
1973	EC 9	• Accession of Denmark, Ireland and Great Britain; after a referendum Norway decided not to adhere	
1976	EC 9	• EC-Foreign Ministers agree at the Hague to create a 200 mile EEZ from 1 January 1977, to which member states' vessels would have free access • The EC acquired exclusive legal competence for international fisheries relations, the third pillar of the CFP, agreements with third countries is settled • Attempt to introduce an EU regime of TACs and quotas fails	100/76 101/76
1976–1983	EC 9	• Negotiation on the conservation policy pillar of the CFP	
1981	EC 10	• Accession of Greece	
1982	EC 10	• Adoption of the United Nations Convention of the Law of the Sea,which introduced the 200 mile EEZ world-wide	
1983	EC 10	• Establishment of the last pillar of the CFP, the conservation and management policy, which introduced a TACs and national quotas system.	170/83
1986	EC 12	• Accession of Spain and Portugal doubled the size of the EC's fleet; restrictions on access to the Community waters until December 31, 2002	
1992	EC 12	• Interim review of the CFP: introduction of a Community system of fishing licenses, adoption of a Community control system for the entire fishing sector; and settlement of a multi-annual framework for several fish stocks.	3760/92
1995	EU 15	• Full accession of Spain and Portugal • Accession of Sweden, Finland and Austria; after a second negative referendum, Norway fails again to join the EU.	
2002	EU 15	• Reform of the CFP: adoption of multi-annual recovery plans for stocks in danger of overfishing; establishment of a fund for encouraging the decommissioning of vessels	

Sources: Coffey 1995, Lequesne 2000a, Symes 1997; Song 1995

Iberian countries doubled the size of the EC's fleet, member states agreed on access restrictions for both countries to Community waters. In 1992 the first reform of the CFP took place by introducing a Community system of fishing licenses, by adopting a Community control system for the entire fishing sector and the setting of a multi-annual framework for several fish species. The fourth enlargement of the EU in 1995 with Austria, Finland and Sweden did not lead to large changes in the CFP, as fisheries are not relevant for the Finnish and Swedish economy. The situation would have been totally different if Norway, a key fishing state *par excellence*, had joined. After a referendum, however, for the second time the Norwegians rejected the adhesion to the EU.[4] Finally, in December 2002 member states after lengthy and difficult negotiations agreed on a second reform of the CFP. The new measures entered into force on January 1, 2003. The main elements of the final compromise package are as follows: the abolition of public aid for the renewal of the fishing fleet after December 31, 2004 and tougher conditions for subsidies for the modernization of boats; the introduction of multi-annual recovery plans for stocks below safe biological limits (for example cod stocks) and management plans for stocks within it; increased premiums for the scrapping of vessels in order to achieve additional reductions in fishing effort resulting from recovery plans for the period 2003–2006. In the coming years with the integration of the Central and Eastern European countries the central issue for the CFP will be how to integrate or accommodate the fleets of countries like Poland and Estonia, which bring with them few fish stocks, but many fishing boats.

This short description of the key dates in the development of the CFP, solely illustrates the present state of the CFP, but does not explain why fisheries came into the agenda of the EC. On the same line of argument as Lequesne (2001) and Symes (1997), it is assumed that the settlement of the CFP may be explained by external and internal policy input variables. While the external policy inputs refer to the adaptation of member states to the changing environment, internal policy inputs deal with changes in the structure of the polity EU itself.

No specific mention of fisheries policy was made in the Treaty of Rome, apart from a remark in the agricultural policy defining fisheries as an "agricultural product". The fisheries sector was a minor issue for the original six member states, who concentrated their attention on the CAP. Furthermore, at that time, nearly 90 percent of the fish catches of Belgium, France, Germany, Italy, and the Netherlands were taken outside their national fishery limits, that is to say outside their 12 miles zone (Holden 1996, 17). At the end of the 1960s, demands for a CFP arose because of external and internal inputs. The internal input was the creation of the EC common customs tariff, an external tariff for products coming from outside the EC. Furthermore, there were two external inputs with which the six member states had to cope: the negotiation of tariff concessions in the GATT framework, and the first enlargement of the EC. The settlement of the common customs tariff and the GATT framework resulted in a liberalization of trade in fish products within and outside the EC making a common market organization for fish products necessary. Before going to negotiations of tariff concessions in the GATT framework, the objective was to create a common market for fish products inside the EC. In order to achieve this, customs barriers and any other measures that could prevent the flow of fish products inside the Community (from one member state to another) had to be dismantled and at the same time common rules for the market of fish had to be set up. Consequently, arrangements concerning the common market in fish products within the EU and implementing trade policy measures affecting fish imports and exports were adopted. Similarly, to the CAP, the Council started to determine annually the fish prices and the import quotas for the following year. Thereby, setting guide prices for fish trade within the EU and introducing reference or minimum import prices for imports from third countries.

Enlargements are a major external input in any assessment of policy transfer, which bring about a shift in the number of participating actors and in the configuration of players' preferences. The CFP constitutes no exception to this rule. Successive enlargements of the EU were a key factor in the inclusion of the fisheries policy on the agenda of the EU and also on its evolution. Since the beginning of the concept of a CFP in the early 1970s, the EU has witnessed three phases of enlargement, each adding new actors to the bargaining process. The number of participants has increased from six in 1957 to fifteen in 1995, and this number will grow further when Central and Eastern European countries are given EU membership. The first enlargement (Great Britain, Ireland, and Denmark) changed the configuration of preferences in the fisheries policy of member states. All the accession candidates were key fishing states, that is to say they all had large fisheries resources and production capacities.

In the late 1970s two external inputs affected the CFP once more: changes in the international fisheries jurisdiction with the establishment of the EEZ, and the beginning of accession negotiations with Portugal and Spain. A major change in the whole approach towards the management of fish stocks came from the negotiations on the UNCLOS, which began in 1973 and culminated in the adoption of this convention in 1982 with the extension of national fishing zones from 12 to 200 mile EEZ, in order to improve the overfishing situation in major fishing stocks.[5] With the subsequent ratification of UNCLOS in 1994, the trend towards expanded fishery jurisdiction was formalized permitting the coastal state to have exclusivity in the management of the majority of fish stocks. This change in the international fisheries jurisdiction had severe consequences for some EC's countries, because the introduction of the 200 mile EEZs restricted considerably the access of the European fleets to fish resources in other parts of the world, especially in the North Atlantic. Great Britain and Germany, which had always fished freely close to the coast of Norway, Iceland, and Canada, found themselves suddenly excluded and were forced to seek exchange deals in order to gain access to those fish stocks. As a direct result of this change their distant water fleets were strongly curtailed. Moreover, being forbidden to fish in the waters of the northern Atlantic made it even more difficult to deal with the increased energy costs after 1974. A vicious circle seemed to emerge: the higher petrol costs could only be matched by larger catches, but those were only possible in colder, northern waters, which were now barred. Fishermen in Canada, Norway and Iceland had the advantage of lower competition, allowing them to make larger catches and to sell them in the European markets at prices below those of European fishermen, who could no longer fish where they had been fishing for decades (Shackleton 1983, 352).

The second external input, the accession negotiations with Portugal and Spain, led to a consensus that it was necessary to agree to a conservation and management policy for fisheries resources before these two key fishing countries joined the EC, since both states had large fishing fleets and a strong preference on having access to the 200 mile EEZ of the EC. When the Iberian countries applied for membership, Spain then had the fourth largest fishing fleet in the world and it was equal in size to three quarters of the total fleet of the ten. After accession, nearly a third of all Community fishermen would be Spanish. Hence, the accession of Portugal and Spain changed once more the configuration of preferences within the EU.

Due to these different external and internal inputs member states had to try to cope with a changing environment and gradually transferred the competence of the fisheries policy to the European level.

COLLECTIVE ACTION AS A MEANS OF REDUCING EXTERNAL COSTS

The policy transfer was motivated by the prospect of mutual gain from coordinated action. It was realized that a common fisheries policy was better than a unilateral effort to respond to the change in the international fisheries regime. In situations where jurisdictions are linked through economic interdependence, different national policies would weaken the effectiveness of each. Or as one actor involved in the negotiations stated:

> "Ich glaube, wir sollten [...] eines nicht vergessen in der so schwierigen Situation der Ausdehnung auf 200 Seemeilen entsteht [durch] eine Gemeinsame Fischereipolitik, eine Position, wo wir wirtschaftlich und politisch enorm in der Gestaltung des Fischereirechts der Welt mitwirken können. Das hätten wir als einzige Mitgliedstaaten kein Land für sich gehabt, sondern wir wären den großen Küstenländern ganz ausschließlich ausgeliefert gewesen, das ist das, was uns beflügeln muß."[6]

The decision to transfer a policy competence to the supranational level is a means of reducing external costs imposed on nations. Buchanan and Tullock (1962, 43–45) distinguish between two kinds of expected costs: external and decision-making costs. Whereas the former are a result of others' actions over which a nation has no direct control; the latter are a result of the participation of one actor in an organized activity (here: EU-membership). The sums of both costs are the so-called interdependence costs.

Interdependent relationships always involve costs, because interdependence restricts autonomy. When analyzing the costs and benefits of an interdependent relationship, two different dimensions have to be taken into account: absolute and relative gains. Whereas in absolute gains the concern is on how well actors (states) fare themselves, relative gains emphasizes how well the actors fare when compared to other actors (cf. Snidal 1991, 703). Moreover, these relative gains are affected by distribution issues from how those gains are divided. It means that costs and benefits of an interdependent relationship, in the analyzed case a common policy, are evaluated taking into consideration how others are doing.[7] Accordingly to Snidal (1991, 703) relative gains inhibit cooperation in two ways: relative gains limit the range of possible cooperative agreements, because bargaining outcomes benefiting disproportionately one or more actors will not be accepted; second, when distribution is the main relative gains problem, actors may change the terms of a cooperative agreement or offer side-payments until a proportionate distribution of gains is given.

An actor finds it advantageous to organize a political activity collectively only when s/he expects to increase its utility. This can be done in two ways. On the one hand, collective action may eliminate external costs of actions, which other actors impose upon the actor in question. On the other hand, collective action allows some additional external benefits to be gained which an actor alone cannot secure (see Buchanan and Tullock 1962, 24, 43–44). In the same line of argument, Moravcsik (1993, 486) distinguishes between two major purposes of policy coordination in the EU. First, it enables the accommodation of economic interdependence through reciprocal market liberalization, as was the case when settling the two first pillars of the CFP, the structural policy and the common market organization. Second, it intends to harmonize policies with the purpose of assuring the continued provision of public goods or, in the case of fisheries, of common pool resources. In public goods theory goods are classified according to two criteria: non-excludability and non-rivalry in consumption. Non-excludability is given when no one can be excluded from using that good. Furthermore, public goods are non-rival or indivisible it means that one person's use or consumption does not diminish the availability of that good. National defense or public parks are some of the best-known examples of public goods.[8] Whereas common pool resource is a natural resource sufficiently large, in which it is costly to exclude users from obtaining subtractable resource units. Hence, the two criteria used to define a common-pool resource are non-excludability and rivalry or divisibility. It means that the resource units (for example a fish stock) acquired by one user are done so at the expense of other potential users. Fishing grounds and pastures constitute typical examples of common-property resources.

The next part illustrates how the open access character of fisheries until the adoption of EEZs led to an overexploitation of fish resources.

THE OPEN ACCESS CHARACTER OF FISHERIES: THE OVEREXPLOITATION
OF RESOURCES

Until the mid-20th century the ocean's waters remained relatively open and with their resources accessible to any interested party. A narrow coastal strip placed under the jurisdiction of the coastal state was the only exception. At that time, the oceans were only divided into two spaces: a *high seas regime* with unrestricted fishing, that is outside a narrow coastal strip all states could fish where and as much as they liked, and a *territorial seas regime* between six and twelve miles, in which coastal states were obliged to share fishing rights only with those states which had traditionally fished there (Garcia and Hayashi 2000, 447).

In fisheries there can be distinguished between two related, but different, concepts: open access and common pool. Because both terms are often used interchangeably, it results in some confusion. Nevertheless, while open access refers to a fishery into which there are no restrictions on entry or exit that stem from ownership of the fishery resource (rivalry in exploitation), suggesting an absence of property rights in the resource *(res nullius)*, for example high seas fishing. The exploitation of open access resources leads to negative externalities for the users of the resources, which causes the tragedy. Unlike a common pool resource is one to which a number of owners have co-equal rights of use, but not to transfer the entire resource *(res communes)*. Thus, an use of a common pool resource means that a set of property rights exists, that the number of users is limited, and that each user respects the established rules and conventions in the sense that s/he knows how much of the resource s/he may extract (Stevenson 1991, 5; Weimer 1997, 3).[9] The open access character of fisheries until the beginning of the 1980s played a significant role in the overexploitation of many of the world's major fish stocks. Demand and catch capacity outgrew the biological capacity of renewal, and lead to a overfishing of resources. World wide, fishing reached its maximum production level one or two decades ago, and now there is a declining trend in total catches. According to a study from the United Nations Food and Agriculture Organization (FAO) (1997, 7–9), almost 44 percent of the major fish stocks are fully exploited with the consequence that catches are very close to their maximum limit, although the pattern varies regionally and by species. About 16 percent of species are overfished and about six percent are regarded as depleted. In some of the worst affected marine fishing regions (the North-West, North-East, West-Central, South-East Atlantic, the Mediterranean and Black Seas and the Eastern Central Pacific) catches have decreased by more than 30 percent. The Atlantic fisheries, with a long history of fishing, have experienced the largest declines, although the Mediterranean and parts of the Pacific have also suffered serious declines. In some fisheries there has been a general collapse of stocks. For example, the cod stocks of the Northwest Atlantic (Canadian Grand Banks), for a long time considered the world's greatest source of fish, collapsed in the late 1980s and since 1992 there is a moratorium on the exploitation of this stock (Ávila de Melo and Alpoim 1999, 8).[10] The herring fishery of the North Sea was also overexploited and closed between 1977 and 1983 in an effort to give the depleted stock a chance to recover.

The successive collapse of the two greatest fisheries of the North Atlantic (North Sea herring and the Grand Banks cod) made obvious that the methods of the fisheries management required drastic modifications. Since the single nations had been unable to enforce control of these shared stocks and due to the increase in overfishing of some fish stocks, the concept of "freedom of seas" and the open access character of fisheries beyond the 12 miles had to be altered and EC member states had to adapt to this external input.

THE TRAGEDY OF THE COMMONS APPLIED TO FISHERIES

The open access nature of the fisheries until the settlement of the EEZ constitutes a typical paradigm of what Hardin (1968) labeled the tragedy of the commons, referring to the dilemma of common interests. In the case analyzed by Hardin the commons were pasture and grazing grounds open to all. This unrestrained individual use resulted in the tragedy, that is to say in over-grazing. This metaphor of the tragedy of the commons is used to illustrate that through the lack of restraint in using the commons, which are by defi-nition owned by everyone, all actors will end up with a Pareto-inferior move, which is worse than any other possible outcome for at least one actor. The so-called tragedy is by no means peculiar to overexploitation in agriculture, as is evident in the case of other common pool resources, such as air, water, and fish stocks. Thus, if the system functions as one of a free-for-all, none of the participants will have an incentive to consider the effect of his effort on the returns to the efforts of others, and in the end this will lead to a complete degradation of the resources. The private benefits of grazing an additional head of cattle on a common range exceed the common costs, because the costs of maintaining range quality can be shifted to the whole group. As common pool resources are free goods for the individual and simultaneously scarce goods for the society, the individual is locked into a system that brings ruin to all (Hardin 1968, 1244).

The ideas sketched above illustrate the problems associated with the use of common pool resources. Fish resources are an excellent example of the problem of sharing a common depletable resource. When the herders in the example mentioned earlier are substituted by fishermen, fishermen seek to maximize their own short-term interests, that is, fishing as often and as hard as they can, the result being the depletion and extinction of many fish stocks. They behave this way, because the fish left behind today are valueless to them, as there is no assurance that they will still be there tomorrow. Each fisherman has a "dominant" strategy to be uncooperative, because there is no incentive to cooperate. In game theory there is the Prisoner's dilemma to illustrate this situation. In this uncooperative situation two prisoners are held in separate cells both suspected of having committed a crime. Both prisoners (or players) are faced with the dilemma of whether to confess or not. Each player knows that if s/he confesses while the other does not, s/he will be pardoned. If both confess, then each of them will receive the full sentence. If neither confesses, then each will receive the shortest sentence. The best outcome for both prisoners, or in this case for the fishermen, is to defect, despite the fact that both would be better off if they were to cooperate, because neither player acting unilaterally has the incentive to deviate. When all players choose their dominant strategy, they will end up with a Pareto-inferior outcome.

In sum, the tragedy of the commons illustrates a situation, in which users, in their efforts to maximize their individual gains, often overuse one resource to the detriment of all. Due to the non-existence of entrance barriers each fisherman worsens the situation by increasing the number of fishing partici-pants while the fish stocks diminish. Consequently, individual rational bei avior in a group leads to less than optimal collective outcomes for ev .ryone. The introduction of the 200 mile EEZ constituted the first attempt to improve the overfishing situation of a common sea. In the EU the decision to settle a conservation and management policy with TACs and national quotas as the principal policy instruments was a way of trying to move from a sub-optimal outcome to one in which all the involved actors (member states) exercise mutual restraint by agreeing in managing the fish resources and consequently by coordinating their strategies in that policy field.

THE STATE OF THE ART IN THE COMMON FISHERIES POLICY

Before turning to the empirical assessment of EU negotiations and before using the conceptual framework for analyzing negotiations in the EU, the scientific literature on the CFP will be briefly summarized here.

A central issue when studying the CFP is to focus on the evaluation of the conservation and management of fish stocks, which is based on TACs and on the national quota system. Although there is a widespread agreement within the scientific community that the conservation and management policy of the EC has not reached its aims of solving the problem of resource scarcity, studies diverge strongly in explaining the causes of this failure.

Some authors (Foders 1994; Gray 1998) explain the poor performance of the CFP with the different set of tools for reducing fishing efforts, such as the TACs, the quota hopping system[11], technical conservation measures, decom-missioning of fishing vessels[12], limited entry legislation, that is restriction of the number of days at sea when vessels are allowed to fish, and enforcement procedures[13]. Other authors (Symes 1996; Cann 1997) outline the gap between management elite and the fishermen, who have no part in the deci-sion-making system. Payne (2000) explains the resource management failure with the nested institutional context of the EU, which led to a political deadlock and the maintenance of the status quo. Another branch of research centers on the description of the settlement and evolution of the CFP (Coffey 1995, 2000; Farnell and Elles 1984; Hatcher 2000; Harnier 1996; Holden 1996; Lequesne 2000a; Symes 1997). There are also some general reports dealing with the situation of the national fisheries industry branch (OECD 1996; Symes and Phillipson 1997), or with fisheries dependent regions (Symes 2000; Freire and García-Allut 2000).

Most of the existing studies, however, are of descriptive or normative character. Because they are not based explicitly on an analytical framework the strength of their explanation is reduced. Single studies based on a theoretical framework have utilized the game theory (Foders 1994) and a three-level game approach (Payne 2000) to explain the conservation failure of the CFP in the EU.

Only a few authors (Leigh 1983; Payne 2000; Symes and Crean 1995) have linked the CFP to the overall European integration process. Coming to the conclusion that it provides an excellent case study in the European integration process, as the study of the CFP tests the willingness of member states to delegate competencies to the European level for the management of a common resource and it allows to analyze how the decision-making process in the EU works in practice.

This short overview of the scientific literature on the CFP is not exhaustive. There also studies focusing on marine and fisheries ecology (Cushing 1995; Mackinson et al. 1997), on bio-economic models of fisheries (Conrad 1996; Sandberg et al. 1998).[14]

What is missing in the study of common policies in the EU are more actor oriented studies. A closer look at the negotiation process itself is necessary, in order to define who are the players sitting at the negotiating table, which preferences do they have, and how far the institutional setting, that is the structure in which the game is played, and the preferences of the Commission matter. By doing so this dissertation attempts to explain bargaining outcomes in EU negotiations through an actor-oriented approach. The CFP will be used to illustrate integrative and distributive bargaining in the EU and whether the actors involved behave differently in the two analyzed bargaining games and how they deal with the unanimity voting rule. In other words, the negotiations on the settlement of the CFP will be used as a means to an end, that is to illustrate how the bargaining game is played in the EU. For that purpose the different bargaining stages will be analyzed, namely the negotiation process from the initial bargaining positions or policy preferences of member states and the institutional setting in action to the bargaining outcomes finally agreed upon.

CHAPTER 2

Constructing a Springboard
A Conceptual Framework
for Analyzing Negotiations in the EU

SETTING THE STAGE

Before starting with the settlement of the conceptual framework for analyzing negotiations in the EU, it is necessary to make a few introductory remarks on its status. A theoretical framework must be specific enough to offer guidance in the field analyzed, yet general enough to permit application to widely variable situations. The idea behind this is to try to create a framework enabling the analysis of the internal negotiations between the Council of the EU and the European Commission. The purpose of this dissertation is not to create an operational bargaining theory to help to understand the specific outcome of any negotiation, but rather to analyze how negotiations take place under those policy fields falling under the competence of the EC, the so-called first pillar[1] of the EU, and also falling under the consultation procedure[2]. Thus, the role played by the European Parliament (EP) will not be taken into account, as the EP only delivers an opinion to the Council of the EU. The power of the EP to influence legislative proposals varies according to the legal basis within the Treaty of Rome. Most fisheries proposals concerning internal fisheries have, as their legal basis, Article 37 of the Treaty establishing EC, which simply requires the EP to be consulted.[3] Although there are several studies available focusing on the EP (see, for example, Tsebelis and Garrett 2000), including this institution in the present analysis would not contribute to a better understanding of the bargaining outcomes and would only make the analysis more complex.

The major function of the conceptual framework is to identify a set of variables that possibly explain the bargaining outcomes. Once these variables are dissected, a closer look will be taken at an integrative and a distributive bargaining game in the fisheries policy, in this way trying to monitor the bargaining behavior of the actors involved. In order to create a conceptual framework for analyzing the EU negotiation process it is important to define the actors, who are able to influence decision-making. The EU negotiation process involves a multitude of actors. Only the role played by the member states and by the European Commission in the negotiation process is considered in the analysis. The member states are represented by ministers or other national government representatives (the principals), whereas the European Commission (the agent) has civil servants, a general director or a commissioner to represent its position at the negotiating table. Each member state is portrayed as a "player" who calculates its benefits both with and without policy coordination.

Negotiations will only be successful if all the member states involved expect their outcomes to be larger with cooperation than without. All actors are utility maximizers, each of them behaves rationally, that is attempts to obtain an outcome as close as possible to its most preferred point. Actors' preferences can be deducted by their negotiating position in a certain policy issue. The member states wish to have a policy reflecting as far as possible their negotiating position and the Commission wishes to expand its power. Finally, the conceptual framework used will be kept simple by assuming that all negotiations are games of perfect information, that is to say the rules of the game and the preferences (negotiating positions) of the players are regarded to be common knowledge. Players have perfect information of the other players' revealed preferences and about the structure of the game. Revealed preferences means that when confronted with two or three alternatives, each actor will vote for the one s/he ranked higher. Moreover, at every stage of the game players are completely informed about previous moves. Uncertainty, however, plays a role in the negotiation process. At the opening bid all the parties involved know that a certain player has transitive preferences about a certain issue, but they do not know whether s/he is going to be able push through his/her preferences. Nevertheless, at some stages of the bargaining game some players have less than full information about the way a player is going to behave in the next bargaining round. It is expected that the *bargaining process itself might matter*, in the sense that players can use the actual negotiation process to bargain strategically with the other players in order to try to get concessions or side-payments as the price to be paid for their approval of a certain policy proposal, which they do not regard as being to their own advantage.

Figure 2 is a graph of the conceptual framework for analyzing negotiations in the EU. It is assumed that the starting-point for negotiations at the Council of the EU, in the figure represented by the abbreviation CEU, is a way of trying to cope with a changing environment, which is represented by the external input (ei). The member states, assumed to be the national governments (G), come to EU negotiations with exogenously given preferences (P MS_1 to P MS_n). Their preferences are defined by their time horizons (TH), the range of interest groups and their organizational effectiveness (in the figure abbreviated as IG), level of politicization and domestic salience of an issue at the national level (LP), and the role played by national parliaments as potential internal veto players (NP). On the other hand, the preferences of agents (P A) also are taken into consideration. The preference of the European Commission (COM) is dissected and it is assumed that its preference formation is that of being a representative of the European Community (REC) and also a rational maximizer bureaucracy (RMB). The preferences of the principals and of the agents interact at the Council before and during the negotiations mainly through the agenda-setting power of the Commission. The Council, however, holds another key intervening variable for explaining bargaining outcomes: the institutional setting (IS), which is constituted by formal and informal rules. Formal rules (FR) refer to unanimity voting (UV) or qualified majority voting (QMV), and informal

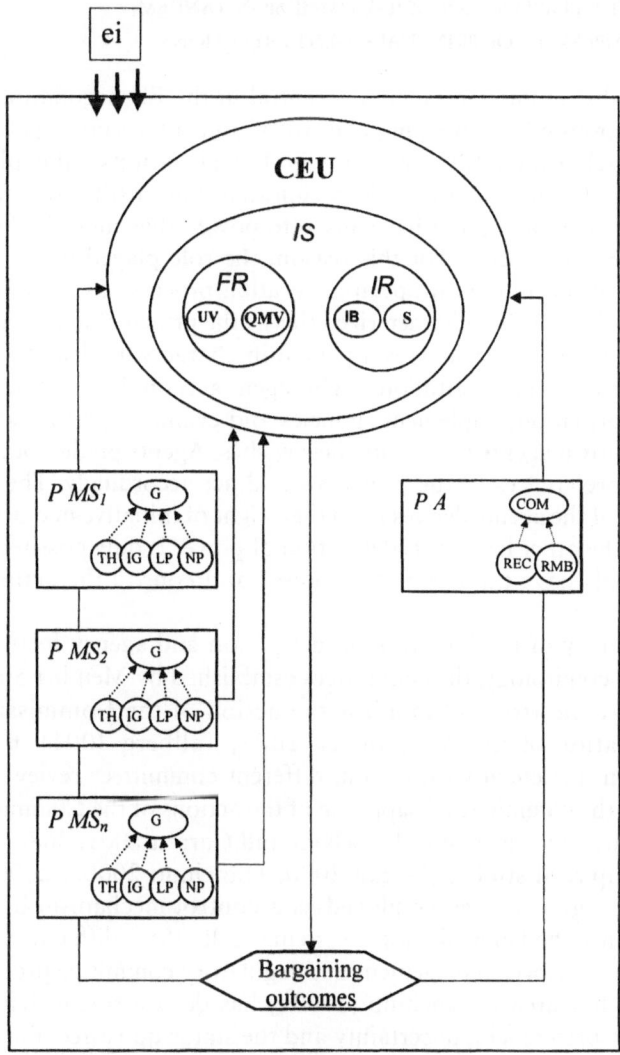

Figure 2: A conceptual framework for analysing negotiations in the European Union

rules (IR) concern the iterated bargaining (IB) and the shadows of the past, present and future looming over the actors (S). These three intervening variables, *preferences of member states, preferences of the Commission,* and the *institutional setting* are expected to explain the bargaining outcomes. Thereby, in the conceptual framework only negotiations under unanimity voting will be considered, and are assumed to be the constant parameter of the conceptual framework. Variation will be given by adding parties to negotiations and by analyzing two different bargaining situations, namely an integrative and a distributive bargaining situation.

DEFINING THE PLAYERS: COUNCIL-COMMISSION TANDEM OR THE COMPLEXITY OF PRINCIPAL-AGENTS RELATIONS

The Council-Commission tandem is central in the EU negotiation process. Both institutions have dominant policy roles[4] and at the same time are dependent on each other. Intergovernmental negotiations take place at a supranational arena surrounded by supranational institutions. Although supranational actors do not have any veto power, they indeed influence the decision-making process. For this reason, the role played by the supranational institutions in the European integration process has to be taken into account. Pollack (1997) characterized this as the principal-agent problem. A principal (EU member states) is a person in charge who delegates tasks to agents (supranational institutions). This agent acts on the principal's behalf, collects information, implements policies and evaluates performance. Principals need loyal agents to execute their wishes. Agents predict or anticipate the policy preferences of the principals and act accordingly. The apparent dominance of the agents derives from their right of initiative in drafting legislation. On the other hand, this delegation of power is only possible, because the principals have the means to monitor and control the actions of the agents.

The majority of the literature on delegation and agency focuses on the question of comitology, the committees established by Member State representatives to control and monitor the action of the Commission in its implementation of EU law (Pollack 2001; Tallberg 2001). Before the Commission implements legislation, different committees review it. In the cases when the committees disapprove of the actions of the Commission, the issue is returned to the Council (Tsebelis and Garrett 2001, 368).

Some empirical studies (Dogan 2000; Franchino 2000a, 2000b) argue that comitology is indeed employed as a control mechanism by member states, because the Council adopts systematically three different committee procedures: advisory, management and regulatory committee procedures in different policy areas. Franchino (2000b) has demonstrated that decision rules, policy preferences, uncertainty and the status quo affect the degree of executive discretion given to the Commission. The degree of control varies according to the applied committee structure.

Under the advisory committee procedure the Commission has to refer its proposed actions to the committee and then decides, taking account of the committee's opinion. This procedure applied mainly in the competition policy, gives the Commission the greatest autonomy and the member states the weakest influence. In the management committee procedure, the Commission submits its implementing measures to the Committee, which may initiate a vote by qualified majority within a deadline set up by the Commission. When the committee gives a favorable opinion or fails to give any opinion before the deadline, the Commission may adopt the measure. If, however, the Commission is opposed by a qualified majority in the committee, then there are two possible variants: the Commission may delay the application of its decision for up to one month or for a period up to three months. Thereby, within these deadlines, the Council of the EU may take a different decision by using a qualified majority vote. Thus this committee procedure is more restrictive than the advisory committee procedure and it is used mainly in the CAP and in the CFP. Finally, the regulatory committee procedure was settled to control the Commission more closely than the management procedure. Under the regulatory committee procedure, the Commission is only able to adopt the measures which were previously approved by qualified majority within the committee. Whereas the management committee procedure requires a qualified majority to secure a reference to the Council, in the regulatory committee procedure a minority is able to secure that reference. This procedure is used for legislation regarding customs, veterinary, plant, health and food matters (Pollack 1997, 115–116; Tallberg 2002, 30).

In general, the actions of the Commission are undisputed only when fulfilling its function as guardian of the treaties, and there is no necessity for control. The existence of these different committee structures allows the member states to maintain their influence over policy implementation and to keep their roles as "gatekeepers" in the EU decision-making process constraining, in this way the autonomy of the Commission.

The Principals: The member states or the ultimate deciders

The member states are represented at the Council of the EU, which is an assembly of national government representatives, namely ministers, secretaries of state, or representatives of the different national ministries. The Council is considered to be the institution of the EU *par excellence*, because it is also the main legislative body of the EU. Not only does the Council carry out a policy- and decision-making function in cooperation with the Commission, but is also a forum for conciliating the distinctive positions of the member states (Hayes-Renshaw and Wallace 1997, 2; Lindberg and Scheingold 1970, 83).

In practice, the Council consists of a pyramid of meetings, which go from the working expert to the ministerial level. The ministers representing their own government are normally accompanied by technical personal, from the COREPER and from the national administrations. In order to prepare and discuss technical proposals submitted by the Commission, the COREPER establishes numerous working groups and other technical committees made up of civil servants. This explains why there is now a complex network of committees through which member states representatives bargain over policy proposals and where national bureaucracies meet each other constantly in a myriad of committees. The most important Councils by far are the General Affairs Council, which is in practice superior to all the other Councils, the Budget Council, the Agriculture Council and the Economics and Finance Council (Hayes-Renshaw and Wallace 1997, 29; Westlake 1995, 59). Because the establishment of the Community EEZ and the guidelines for the CFP were considered to be of wide political importance, foreign ministers initially discussed fisheries issues at the General Affairs Council. The first Fisheries Council was introduced in 1977.

According to state-centric theoretical approaches the national government representatives perform the role of prime "gate-keepers" in the EU. They aggregate national level demands and transmit them to the supranational institutions. Furthermore, state-centric theories have a clear conception of the relationships among the primary policy actors in the EU. Whereas the relations among member states are based on formal equality, the relation between the Council and Commission is strictly hierarchical. It is without question that member states play a central role in EU negotiations, but they do not have full control over these negotiations. State-centric approaches neglect the gradually unfolding implications of shared decision making, since their focus is limited to intergovernmental bargains. Although member states are extremely powerful, the path to European integration has placed the principals in a complex institutional setting, which can not be explained with interstate bargaining only. When trying to give an accurate picture of bargaining outcomes in the EU, the preferences of the Commission and the institutional setting also have to be taken into consideration.

The European Commission: A simple agent of member states?

The theory of intergovernmental bargain recognizes that representatives of the Commission can be seen rather as "independent figures than instructed agents" (Keohane and Hoffman 1990, 281), paradoxically at the same time it emphasizes that member states still play the dominant role in the decision-making process, explaining this with the existence of innumerable committees of national and Communitarian experts. For Moravcsik (1993, 507) the Commission is a mere international institution cementing the existing intergovernmental bargains. The liberal intergovernmentalist perspective assumes that supranational institutions strengthen the power of member states in two ways, first by increasing the efficiency of interstate bargaining and second strengthening national political leaders *vis-à-vis* interest groups at the domestic level. According to this view supranational institutions are the result of strategic calculations of member states. In state-centric theoretical approaches their role is regarded as marginal, involving only lowering the costs associated with the implementation, monitoring and enforcement of intergovernmental decisions.

In the last decade, political scientists have begun to use concepts from the new institutional economics in an effort to explore the hierarchical structure of international organizations. In European studies, the new institutionalism uses a principal-agent framework to demonstrate the tension existing between autonomy of supranational institutions and how national government representatives are able to exercise control over their supranational agents. Principals delegate authority for various reasons. The problem lies in the fact that delegation is also linked to potential costs, arising when the preferences of the principals and of the agents do not coincide. In this situation, agents behave opportunistically (*agency shirking*). Although principals contract agents to perform specific functions on their behalf, the agents may remain independent actors with their own preferences. They often attain more information than the principals and also acquire role-specific expertise in the course of the contracting relationship (Doleys 2000, 537). It has been demonstrated that the Commission might be able to introduce its own preferences into the negotiation process. In some policy areas, such as competition policy, Schmidt (1997) has argued that the Commission has pursued an active strategy of using its powers to introduce controversial policies such as the liberalization of national monopolies. When analyzing the specific role of supranational institutions in negotiations, Pollack (1997) found that the role of the Commission is variable. The Commission's autonomy is greatest when it is supported by diverging member state government preferences, alliances forged with interest groups, or pro-integrative European Court of Justice decisions. Other studies (Marks *et al.* 1995, Pollack 1997) point out that the successive extension of qualified majority voting (QMV) through the different Treaty reforms has strengthened the Commission's role as a broker and as a negotiator, as a minimum of two large member states and one small one (with the exception of Luxembourg) is needed to block a proposal.

According to Thatcher and Stone Sweet (2002, 4) agents fulfil four functions: they facilitate commitment problems, reduce information asymmetries, enhance the efficiency of coming to decisions, and take blame for unpopular policies. These four functions can also be applied to the European Commission, since it has been entrusted with enforcement powers in order to encourage policy commitments. Second, the Commission's initiative and execution power is inter-linked with the development of policy-relevant expertise. Third, the delegation of power to supranational institutions pursues the objective of increasing the efficiency of EU decision-making and allows politicians to save time. Finally, the close involvement of the Commission in the decision-making process allows national government representatives to blame agents for policy failures (cf. Tallberg 2002, 26–27).

The Commission also functions as a monitor, that is, it insures that contractual terms (treaties, directives and regulations) are respected (Doleys 2000, 541). As Hosli (1995, 66) stated, supranational institutions have an important function in coordinating the behavior of actors, in providing negotiation structures in which the collective outcomes are Pareto-superior to a situation without policy coordination. Furthermore, the Commission must behave neutrally and treat all member states equally. It also coordinates the available information, to enable the exchange of different views regarding policy positions.

Conversely to Bueno de Mesquita's (1994) model, in which the focus was on how member states bargain among themselves, excluding the role played by the European Commission. In this book, the argument is that it is not simply an agent, but an actor with its own preferences. In the political system of the EU the Commission is provided with a certain decisional autonomy at the negotiating table almost equal to that of the member states. The Commission, the "chief policy-formulating body" (Wallace 1978), sits at the Council with the member states as a "virtual bargaining equal" (Lindberg and Scheingold 1970). This explains why some authors define it as the "sixteenth member state" (Hayes-Renshaw and Wallace 1997) within the Council under the first pillar. Officials of the Commission (commissioner or the general-director) always take part in the bargaining process in order to defend the policy proposed by the Commission at the negotiating table.

Although the Commission has not the power of vote, it has indeed the legislative power of initiative under the first pillar of EU governance and has the command of technical expertise necessary for new policy proposals. In contrast to a principal, it has no veto power and cannot block a decision. The bargaining power of the Commission in EU negotiations stems from its agenda-setting power. The formal agenda-setting power of the Commission comes from the sole right of initiative, allowing the Commission to influence policy outputs. Moreover, through its informal agenda-setting power, the Commission has the ability to exploit situations where national government representatives are confronted with a collective action problem, but avoid further action because of the possible transaction costs (Pollack 1997; Schmidt 1997; Garrett and Tsebelis 1996). A collective action problem is given when individual incentives lead to inefficient collective outcomes. These kinds of problems can arise in two types of situations: common pool resources and public goods. Whereas in the first type, a tragedy of the commons situation is given, that is to say, a group of people has access to a common pool of resources that may be depleted by any member of the group. The second type of collective action problem occurs when a state, an international organization or a regime provide public goods which are by definition indivisible and non-excludable. The overexploitation of fish stocks constitutes an example of the overuse of a common pool resource, and free trade is an important example of a public good in an international regime, the World Trade Organization.

The settlement of international institutions for solving collective action problems leads to an increase in the number of involved actors at the bargaining table. In the political system of the EU, the Commission is not purely a passive tool of the member states, since it has its own preference and financial resources.

PREFERENCES OF MEMBER STATES

Preferences of national government representatives are one of the crucial variables in the analysis of the EU negotiation process. Preferences of actors are relevant, because they motivate the domestic and international interactions. The definition of national preferences is difficult, since they are defined differently on different issues, at different times and by different governmental units. States with a centralized political tradition (such as France) are better placed to define their national preferences than federal or regionalized states, such as Germany or Belgium, with a larger number of divergent actors.[5]

One of the main problems in European studies literature is that the concept of preferences is mentioned, but not further conceptualized. The central question here is how the preferences of member states emerge. The emergence of preferences has long been considered as taking place in a black box. In contrast, Moravcsik (1993, 1997) pointed out that governments define their preferences through the interaction of interest groups.

There is now a broad discussion in the new institutionalist literature on the question whether preferences given are exogenous or endogenous. While sociological and historical institutionalisms assume that they are endogenous, rational institutionalism takes preferences to be exogenous. The three different branches of new institutionalism, however, do not take the step further of trying to define the content of preferences empirically. Hosli (1996, 256) pointed out that the preferences of the actors involved in the bargaining process are either not really known, not shown openly, or not stable during the bargaining process. Garrett and Tsebelis (1996) also argue in the same direction emphasizing that analyses based *a priori* on the preference distribution of member states include many speculative elements. It may be possible that strategic or sophisticated voting occurs, where one actor votes against his/her true preferences with the aim of achieving an even better outcome. For the sake of simplicity, however, it is assumed that member states vote sincerely, that is they vote according to their true preferences. Only the "revealed preferences" at one stage of the negotiation process shall be taken into account. The analytical approach used is based on the rational-choice tradition that actors are assumed to be rational in the sense, that they pursue certain strategies for maximizing their exogenously given preferences. The assumption of stable preferences prevents the scientist from surrendering to the temptation of postulating the required shifts in preferences when s/he wishes to explain all apparent contradictions to his/her model.

Furthermore, the institutional setting, constituted by the formal and informal rules of the game, alters the costs and benefits of various strategies used by actors. The institutional setting, however, does not affect the actors' underlying preferences—that is, they remain stable. When referring to the preferences of member states, this means that actors prefer one alternative to another over certain outcomes. Every actor is supposed to consider the consequences of all the possible options from which a choice could be made, and rank these outcomes in order of preferences, from the most preferred, down to the least preferred. This ranking of preferences, also called preference ordering, must be consistent in the sense that if A is preferred to B and B is preferred to C, then A must also be preferred to C. The preferences of the actors become apparent through their policy or negotiating positions in different issues. Thereby, each actor will try to push through his/her own preferred outcome.

If preferences are taken as the major determinant of the actions of national government representatives, the factors determining the formation of national preferences have to be specified. At the negotiating table the considerations of national politics motivate the behavior of the national government representatives, who have to be concerned about the impact of any decisions made at the supranational level on their domestic constituencies, and also about the impact of those decisions on any future elections (cf. Peters 1992, 79). So far only few analyses consider this domestic arena (Milner 1997; König and Pöter 2001). The conceptual framework developed in this work uses four indicators to define the structure of national governments' preferences: time horizons of decision-makers, interests groups and their organizational effectiveness in one economic sector, the level of politicization of one issue and national parliaments. Time horizons refers to the fact that national government representatives are most interested in the short-term consequences of their actions. The role played by interest groups depends on their organizational effectiveness in mobilizing and influencing the negotiating position of a national government representative and on the distribution of interests within an economic sector. The third factor, the level of politicization, refers to the importance of an economic sector and consequently, to the high or low salience of an issue. Finally, national parliaments as internal veto players will be used as the last indicator for defining member states preferences, since national government representatives (ministers) are accountable to their national parliaments.

Time horizons of decision-makers

When trying to define the structure of preferences of member states, the time horizon of political decision-makers (chiefs of government or other national government representatives) is a crucial factor. Although many implications of political decisions only have an effect in the long run, political decision makers are often most interested in the short-term consequences of their actions. Electoral concerns may explain the behavior of political actors. As Keynes once noted in the long run we are all dead; for politicians, electoral death can come very quickly. The decisions of voters, taken in the short run, determine political success (Pierson 2000, 479). Governments choose those policies that best serve their immediate electoral interests. This explains why politicians prefer to avoid unpopular measures, at least in the second part of their mandates.

Public choice scientists explain time horizons of governments with the argument that politicians are "no better, but also no worse than businessmen" (Tullock 1979, 31). This implies that politicians are more interested in their own than in the collective utility. Milner (1997, 34–35) has drawn our attention to the fact that when a certain government wishes to maximize its chances of reelection, it has to worry about the overall economy and the preferences of interest groups supporting it. This assumption of an "office-seeking" motivation was already introduced by Downs (1957). It means essentially that the policy preferences of governments do not necessarily follow their party platforms nor their campaign promises, but rather that policy choices are influenced by electoral considerations. The utility function of politicians is primarily to maximize their chances of re-election. Short-term outcomes are essential to politicians, as in the next elections they will have to show what they have achieved (Pierson 1996, 147). As elections take place every four years, longer-term objectives are simply banished.

In sum, as assumed by Machiavellian politics, strategies can determine the survival of politicians, whose principal interest is to remain in power and maximize their margin of maneuver. Optimal solutions in a certain policy field are secondary, as they would be too expensive in the short-run, that is would cost too many votes.

Interest groups: Organizational effectiveness and distribution of interests

The variety and effectiveness of interest groups within a sector is a further indicator influencing the preferences of national government representatives. National government representatives do not exist in a vacuum, to use Milner's (1997) words, that is they can not ignore the demands of national interest groups or other societal actors.

The role played by interest groups in the preference formation of national government representatives is given by the organizational effectiveness and the distribution of interests in a certain economic sector. While organiza-

tional effectiveness refers to the ability of interest groups to get their preferences taken into account in a certain policy proposal, the distribution of
interests within a certain sector explains how strongly or weakly organized
a certain interest group is. The more numerous the interest groups are, the
more difficult it should be to aggregate all the interests involved into a policy
proposal. Heterogeneity in a sector affects the form of sectoral interest aggregation at the national level. As Beyme (1980, 105–108) has pointed out,
without an effective organization it is impossible for interest groups to get
their preferences taken into account in the legislative process.

Moravcsik (1993, 484, 505) assumes that the activity of groups is determined by their possible gains and losses. Groups with much to gain or lose
will put pressure on governments, while marginally affected groups will
remain passive. Losers tend to generate more political pressure than winners.
When adjustment costs for domestic groups, who have to adapt to supranational legislation, are high, they have to be compensated. National interest
formation is very much the result of a government striking a balance between
the different pressure groups, which are active in a certain policy field. The
stronger the interest groups are, the smaller the room for chiefs of government to maneuver.

What is missing in Moravcsik's argument is how to explain a limited influence of certain interest groups at the supranational level. So with the transfer
of one competence, one would expect that the power would also be shifted
to the new level of government. Or from an institutionalist perspective, as
Risse-Kappen (1996, 66) emphasizes, the more integrated a policy issue is,
the more one would expect that transnational interest groups start to
develop. The question arises why do interest groups not react to a change;
are they not reacting because of the continued importance of national
governments at the Council or is the non-participation rather an expression
of the wish to save their energies for action at the national level?

In the fisheries policy, *Europêche* is the only body at the supranational
level and was set up by fishing vessel owners with the objective of getting
direct access to the Commission. This body can be compared only in form
to the farmers' organization *Comité des Organisations Professionnelles Agri-
coles (COPA)*, because in practice its structure is rather different. Whereas
COPA has about forty employees, *Europêche* is a European federation of
fifteen national fisheries associations with only a director and a secretary at
its disposal to coordinate its work. In reality, however, not all national fisheries organizations are part of it, because some of them, like the Finnish,
Greek or Portuguese, are not able to pay their contributions (Lequesne 2001,
80). This explains why *Europêche* is often not capable of defining common
positions and of putting pressure on the European Commission (Shackelton
1983, 357; Lequesne 2000a, 353). The strong geographical and sector fragmentation makes it difficult to formulate the fisheries interests at the
supranational level under the umbrella of a single organization. Since the
national fishing industries are regionally and sectorally organized, there is a

fragmentation of their political representation at the national level.[6] In such a situation, it is even more difficult to articulate the interests of the different national fishing industries at the European level.

Level of politicization of one issue

The level of politicization of one issue is defined through its domestic salience. When analyzing a member state's negotiating position in a certain issue it is important to find out if a certain economic sector is important and how sensitive the issue is. The policy position of an actor at the negotiating table will depend on the degree to which an issue is relevant for him/her. Domestic salience of an issue reflects the priorities that each member state attributes to a particular policy issue, which for her/him can have a high or a low salience. When the salience is high (the issue is particularly important) the actor will use all the resources available to try to get the preferred outcome. In contrast, if the salience is low (the member state has no fundamental interest in one issue), the actor is willing to compromise more easily (Arregui 2001, 6).

Fligstein and McNichol (1998, 78–79) define the salience of policy issues according to a high or low ration. According to them from seventeen policy issues, only four can still be predominantly considered as intergovernmental and high ration issues. Those are financial and institutional matters, taxation, fisheries, and structural funds. Curiously, Fligstein and McNichol used the number of adopted legislation and transnational groups in one given issue as criteria to measure the importance of one policy issue. However, this division is intriguing for several reasons. On the one hand, the member states have kept control over the internal financial and institutional arrangements of the EU, issues regarding harmonization of taxes, the fixing and spending of structural funds. On the other hand, member states have agreed to supranational cooperation over a wide range of important issues including monetary policy, agriculture, the internal market, and competition policy.

The negotiating positions of member states are easy to explain when the general context of the interests represented by each state is recognized. When analyzing a policy field in the EU it is important to find out, for which actors a certain issue has a high or a low salience. Thus, in the case of fisheries policy the key fishing states have to be defined, namely Spain, Portugal, Great Britain, France, Ireland, the Netherlands, and Denmark. As a result, fisheries are considered to have a high salience for these countries.

National parliaments as internal veto players

National Parliaments are used as a strategic arena in which governments compete with opposition parties for votes. Thereby the opposition can easily blame the government position for giving up national interests in a specific negotiation in the EU. Before entering EU negotiations, national government representatives are already involved in a whole range of *internal nested games* (Tsebelis 1990), which determine their negotiating position. National government representatives try to settle their preferences through using these two arenas, in which they face different constraints from each side.

When national governments (a majority party or a coalition of parties) make use of their veto power in EU negotiations, two-level games approaches (Putnam 1988) are required in order to try to explain how and why an actor blocks an agreement. There is now a broad range of studies formalizing and extending Putnam's conjecture that domestic constraints can be a bargaining advantage in international negotiations (Iida 1993; Mayer 1992; Milner and Rosendorff 1997; Mo 1994, 1995; Rector 2001). All these two-level approaches echoing Putnam, assume that a negotiator's domestic constraints can be captured by the concept of "win-set", that is, the range of agreements at the international level acceptable at the domestic level.

National parliaments, considered as internal veto players, have to be taken into account, because ministers are accountable to their national parliaments. How strong or how weak the power of these parliaments is, depends on the constitution of each country. This may be illustrated briefly. In Germany, the upper chamber of Parliament (*Bundesrat*) has effective veto power over integrative measures affecting provincial (*Länder*) authority. In Denmark, the European Affairs Committee (EAC) of the national parliament is a strong external veto player in any EU negotiation, because on EU questions it has the function of parliamentary control over the government. Before important negotiations take place at the European level, the EAC has to be informed about the negotiating position of the executive, in order to authorize the government with a negotiating mandate (Nannestad 1997, 61; Pahre 1997, 153). The right to refuse the negotiating mandate has an effect on European policy-making, because *ex-ante* vetoes reduce the flexibility of governments in negotiations.

PREFERENCES OF SUPRANATIONAL ACTORS: THE COMMISSION'S PREFERENCES

Until now, there are few studies (Friis 1997; Hooghe 2001) trying to conceptualize preference formation of the European Commission. Friis (1997) argued that the formation of the Commission's preference is determined by a triple-pressure-structure: first, its own general interest in acting as a motor for more integration; second, the special interests of its own bureaucracy; and the interests of the member states. Hooghe (2001) determined preferences by interviewing over hundred top Commission officials. According to the results of her study, sources of influence in the Commission's preference formation are the socialization (length of service in the Commission), a hybrid category (like the Delors factor), and utility maximization (such as career prospect). Nevertheless, this kind of study based solely on elite interviews, that is on empirical rather than in deductive assumptions, is problematic. Indeed, it is rather doubtful that questions of the kind "what are the reasons for your initial interest in European affairs?" or "what do you consider your main accomplishments and what are your greatest disappointments?" can reveal the preference formation in the Commission. Studies relying on expert interviews can be useful to give the scientist information on the negotiating positions of the actors on a specific issue, but not to try to define empirically the preference formation of a certain actor.

Since the Commission is also considered to be an actor in EU-negotiations, its preferences will be dissected in this dissertation. The preference of the Commission is not only to expand the scope of Community competence to new policy issues, but also to increase its own influence within it (see also Pollack 1994; Cram 2001; Peters 1992). Although the transfer of competencies is important, the decisive factor is to introduce policy instruments that can be monitored at the supranational level. The Commission's preference formation is influenced by two distinct factors, namely by being a representative of the European Community and by its own utility maximizing considerations, which will be explained in the following.

The Commission as the representative of the European Community

The Commission regards itself as the representative of the European Community in being the only actor having legitimacy as an advocate of the interests of the EC. The question arises, however, what are the interests of the EC? It is assumed that the interests of the EC encompass not only the preferences of all member states, but also deepening the European integration process. The Commission can often use the so-called interests of the EC in order to push through its own preferences. Sometimes it is enough to have two or three member states wishing to have a certain policy proposal initiated. In a concrete negotiating situation, at the drafting stage of a proposal the Commission has to try to take the preferences of all member states into account, because this increases the chance of their acceptance of a specific proposal. The Commission tends to embody, select and publicize particular paths on which all actors are able to coordinate. It may provide a focal point at which the preferences of actors converge. Although the preferences of supranational institutions are not identical to those of national government representatives, they have to attend to member states' preferences to conceal their own goals (Koremenos et al. 2001, 1078).

The Commission's function in representing the interests of the Community is to try to find a compromise solution between its original proposal and what is feasible to reach agreement in the Council (Armstrong and Bulmer 1998, 74). Thereby it has to keep a neutral position towards member states, because as soon as it favors systematically some member state(s), the disadvantaged countries will change their strategy and veto decisions in the European arena, or create stable blocking minority coalitions (Eichener 2000, 167). The Commission's duty is to be "the engine of integration" (Nugent 1995) by taking over the roles of a dynamic motor, a broker and a producer of consensus (Wessels 1994, 316). Thus the Commission is not only one of the actors at the negotiating table, but when representing the interests of the EC, may act essentially as a mediator between the different preferences of the member states.

The Commission as a rational maximizer bureaucracy

The general interest of the Commission consists in strengthening and deepening the European integration process. This assumption can be explained by the specific interest of the Commission in maximizing its utility, that is to obtain more power, which can be achieved through more integration or the transfer of competencies to the supranational level (see also Cram 2001, 772) and thus to act independently from the principals. Schneider and Werle (1989, 415) point out that public institutions are characterized by dual features: they are set up to solve collective problems; hence acting as a rule system. On the other hand, institutions do not remain as empty boxes; once settled they start defending their own preferences, trying to maximize their power. One can go even one step further and assume that the Commission acts rationally as a "purposeful opportunist" (Cram 1994) using every window of opportunity to expand its policy domains and to create new ones.

Supranational institutions are not simply neutral arenas, but are actors with their own preferences. Niskanen (1971) was the first to introduce a model of bureaucratic behavior. Against the prevalent view that bureaucrats serve the public interest, he conceived the bureaucrat as a manager, running a division of a private firm, assuming that bureaucrats act to maximize their budget. As Shepsle and Bonchek (1997, 348) emphasize, it does not even matter *why* a bureaucrat wishes to have a big budget, what matters is *that* s/he wants to have increased resources. Transposing this into the Commission means that this actor acts in order to maximize its preference. In effect, the Commission is not a disinterested party, but an actor with its own interests, a "self-motivated autonomous actor" (Smyrl 1998).

The Commission is in the Weberian sense[7] a bureaucratic organization, defined through three features: continuity, expertise and informational asymmetry. By continuity is meant that, unlike member states, the Commission has a long time horizon. Although the commissioners come and go (every five years), the other civil servants (general directors, heads of unit, technical personal) remain longer in office than national government representatives and are free from popular mandates. As Ostrom (1995, 176) emphasizes, institutions last a long time while the interests of political actors may change more rapidly.

Expertise monopoly and the knowledge of every detail gives the bureaucracy power, as it enjoys an informational asymmetry or advantage *vis-à-vis* the principals. The huge formal and informal knowledge of correlations and backgrounds allows public administration to be in the lead of information and action. Like a conventional bureaucracy, the Commission uses its knowledge strategically when it wishes to reach certain goals at different levels of action. The Commission can be considered to be an information nodal point, which receives and processes data. Thereby, the Commission wants to avoid that too much of this knowledge leaks out, because that would limit its possibilities of action (cf. Franke 1989, 162).

These three aspects (continuity, control of expertise, and informational asymmetry) give the Commission, as an institution, power to manipulate the agenda and the negotiating positions of national government representatives. Because bureaucracies know more than their principals about the work they have been "contracted" to do, information is asymmetrical. In such a situation, it is indispensable for principals to monitor through the Comitology procedure and if necessary impose sanctions on agents to ensure that they comply with the conditions of contract. Unless this is done effectively, agents can pursue interests other than those specified by the principals. The asymmetrical information distribution in favor of the agents causes essentially two problems to the principal: adverse selection and moral hazard. Whereas adverse selection, which is also called *ex-ante* opportunism or hidden information, takes place whenever the principal is not sure that s/he is choosing the agent with the most appropriate skills or preferences; moral hazard, also known as *ex-post* opportunism or hidden action, occurs when the agent's actions can not be perfectly monitored by the principals. This kind of problems could be avoided by fixing the agreement between the principal and the agent in perfect, complete contracts. Moreover, such contracts are very demanding and as actors act under bounded rationality, an actual contract will always be incomplete (cf. Gilardi 2001, 4; Milgrom and Roberts 1992, 127–129). In a principal-agent relationship, principals will always try to minimize agency losses linked to adverse selection and moral hazard. In the analysis of the two bargaining situations the way principals try to cope with these agency losses will be taken into consideration, in order to demonstrate how agency costs work in practice and which measures are taken to contain or restrain agency losses.

In a recent study Majone (2001, 113) contests the centrality of adverse selection and moral hazard in the analysis of the European integration process. According to him the principal-agent relationship should be analyzed as based on a fiduciary principle, in which the fiduciary acts in the interest of his client, driven by responsibility, and by his/her own preferences. The transfer of some political property rights, that is rights to "exercise public authority" in a certain policy area, leads independent agencies to act as fiduciaries. This argument, however, neglects the fact that the central advantage is that the Commission does not have one single client, but fifteen different ones. Thus it can decide in which interest it shall act and, whenever some of clients prefer more integration than less, it can act in their name. Furthermore, one of the direct consequences of delegating power to an agent is that this particular agent becomes another player in the game.

THE INSTITUTIONAL SETTING: FORMAL AND INFORMAL RULES

When analyzing EU negotiations the variable institutional setting also has to be taken into account, since the structure in which a game is played affects the scope of the bargaining outcome. It will be distinguished between formal and informal rules. While formal rules refer to the unanimity voting procedure, informal rules concern the iterated aspect of EU negotiations and consequently the importance of the time shadows looming over the actors. Formal rules may influence bargaining outcomes since they are an important determinant of the capacity of a certain political system to aggregate preferences.

Formal rules: the unanimity voting procedure

In the political system of the EU, legislative decision-making procedures involve a Commission proposal, a Council decision, and also the consultation, information or codecision[8] of the European Parliament. Voting procedures in the Council can take three different variants: simple majority (currently eight out of fifteen votes used only for procedural decisions); qualified majority voting (QMV), meaning that a proposal must get at least 62 out of 87 votes (i.e. a 71 percent majority) to pass, and unanimity voting, that is any EU government can use its veto power to stop the EU adopting new law or policy decisions.[9] As simple majority voting is rarely used, for the sake of simplicity, the focus will be only on qualified majority and unanimity voting. After a brief comparison between both voting procedures, unanimity voting will be used in the conceptual framework for analyzing negotiations in the EU as the constant parameter under two different bargaining situations.

Until the Single European Market in July 1987 came into force, unanimity was the norm. In order to achieve the internal market program by December 1992, the member states decided to introduce QMV in a few policy areas interlinked to the internal market. According to König and Bräuninger (1998) the Southern enlargement was a further factor that lead to a change of decision rules at the Council.[10] Since then the tendency has been to extend QMV, and presently the EU makes about two-thirds of its decisions via this mechanism. Nevertheless unanimity is still required in several areas, in which national sensitivities are rather high, such as constitutional issues (revision of EU treaties), admission of new members, social security, allocation of structural and cohesion funds. The question arises what are the advantages and problems from abandoning unanimity voting and moving to QMV.

Hosli (1996) has shown that under QMV the relative influence of member states and of coalitions within the Council has diminished considerably with the enlargements. In a further study using the Banzhaf index, which indicates the potential benefits or costs for a country from a move to QMV in the Council, Colomer and Hosli (2000) have demonstrated that abandoning the unanimity rule in collective decision-making in the EU is beneficial to larger countries, since under QMV they have more votes than smaller states. This explains why small member states are so reluctant to accept an extension of QMV. A further problem, as Laruelle and Widgren (1998) point out, is that the current distribution of votes among member states under QMV does not lead to a fair distribution of power, as the principle "one man one vote" is not respected. Another aspect pointed out by Elgström and Jönsson (2000) is that under the unanimity rule a higher incentive for cooperative negotiations is given. When only a majority is required, the minority can be defeated through overriding it and hence it is not necessary to look for a consensual solution. In other words, QMV involves the risk that some players are bound to lose from it. Consequently, those losing actors might be less willing to cooperate, that is, to accept an agreement against their preferences. In this situation, they can try to create a coalition with other dissatisfied actors in order to block a decision, although such is more difficult than under unanimity voting, where each actor is a potential veto player. Under QMV the role played by side-payments is expected to decrease, as the reluctant actors can be overridden.

Unanimous decision-making is still so important in the EU, because it is the decision rule that prevents one member state being outvoted in the Council. Collective decisions that need the voluntary agreement of all participants meet the welfare-theoretical criterion of Pareto-optimality and it protects actors from being coerced by others. Difficulties will arise, however, with the transaction costs and the disruptive consequences of "strategic voting", when the players conceal their preferences with the aim to getting concessions (Mueller 1995, 50–51). Translating this into the negotiation game in the EU means that as soon as the number of players exceeds a certain number, the costs of reaching collective decisions will be prohibitive.

According to Tsebelis and Garrett (2001, 371), unanimity voting has two effects on the Council. First, the sovereignty of the individual member states is respected and second, it incapacitates the Council in its function as a collective actor. Hence there are essentially two problems associated with the unanimity norm: it is extremely time-consuming, each and every member state can use its veto power during the negotiations, and unanimity favors the status quo. Under unanimity voting any actor can block any agreement until a compromise formula is found with which s/he is satisfied. Every actor is a potential veto player, defined by Tsebelis (2002, 36) as a player whose consent is needed for a change of the status quo. This means that in areas requiring unanimity aggregation capacity reaches its minimum. Or as Elgström and Jönsson (2000, 690) point out unanimous decision-making leads to agreements where the most reluctant actor can determine the level and scope of policy coordination. Hence there is no incentive for an actor to vote for an agreement (policy measures or policy instruments) below his/her own Best Alternative to a Negotiated Agreement (BATNA). Only through side-payments and package deals can reluctant actors be moved to change their positions.

In chapters 3 and 4, in which actors' preferences and the institutional setting in action are analyzed, this issue will be dealt with once more but from an empirical perspective.

Informal rules: iterated bargaining and the shadows looming over the actors

Another variable when assessing negotiations in the EU is iterated bargaining. Iterated or repeated games are games where the same players meet to play a game more than once. Through repetition new possibilities for new outcomes are created, some of which may be far more attractive than the outcome of the game played once. Repeated games have more strategies than games played only once.

Iterated bargaining means that the game (negotiation) is not a one-shot exercise, but a repeated game, in which the interaction between actors will continue in the future and be influenced by present and past choices. *Iterated bargaining might matter* in the sense that actors have to take into account not only the short-time consequences of their choices, but also the effect of those choices on the long-term perspective. The interactions within the system itself constrain the bargaining strategies and the options of actors. The effect of past negotiating positions is strongest when actors repeatedly meet the other actors involved in previous games. Iterated or repetitive bargaining induces actors to be concerned about their reputation, leading consequently to more cooperation than single-shot bargaining (Raiffa 1982, 13). Moreover, when repetition is at play, a negotiator is more inclined to try to establish a reputation for being tenacious and designed for the long-term. Member states might develop a reputation by playing repeatedly in the same way in the expectation that other member states will finally expect them to play this way. The time interval between successive repetitions of the bargaining game will be assumed to be small so that time can be treated as a continuous variable. Each new bargaining round will not be treated by the players as a new problem to be resolved, but it is affected by the foregoing bargaining stage.

In the EU negotiation process, various shadows influence the behavior of the actors, because they know that their relationship is a durable one. Friis and Murphy (1999, 215) distinguish between three different shadows in the ongoing EU negotiation process: the shadow of the past, present, and future. The shadow of the past means that at the Council no negotiation can start from scratch, since choices of the past created consequences for the present and can affect the feasibility of the options chosen in the present and in the future. This is what historical institutionalists call a path dependency (Pierson 1996).[11] Actors become locked through temporal processes into the iterated negotiation system, developing a number of independent norms, rules, procedures and logics that they can not entirely control. As time unfolds, the probability of continuing along the same path increases, while the probability of significantly deviating from the established path decreases. A path dependent sequence of political changes means that it is tied to previous decisions and existing institutions. As the integration process advances, the number of alternatives and opportunities is narrowed down. The member states find themselves locked in a system that they themselves have created by bargaining. The EU is also characterized by a shadow of the present, that is, many different games are being played out at the same time. Finally, the EU is a negotiation system without a clear exit option: the shadow of the future looms over the actors as they become aware that their relationship is a durable one, tying their hands for the future (Héritier 1996, 17). Axelrod (1984) has shown how the game theory can be used to explain the way stable cooperation patterns among individuals with conflicting interests can evolve and that in the situation of the prisoner's dilemma, it is the simple concern about the future that fosters cooperation. In his study of repeated games with the reward matrix of a prisoner's dilemma game, a tit-for-tat strategy proved to predominate. When the shadow of the future looms over the players, they may be more inclined to cooperate in the present encounter. This is based on the assumption that when repetitions of a negotiating game are foreseen for the future, every move is calculated with reference to the opportunity costs associated with the following interaction (cf. Friis 1997, 81; Hayes-Renshaw and Wallace 1997, 265). This fits the EU negotiation process exactly, where member states always have to consider how their positions or actions in certain issues are going to influence future negotiations. When iterating the game shifts the focus from strategies *within* the game to strategies *between* linked games, what in game theory is called "supergame strategies". The existence of these various shadows creates a complex, multi-functional negotiation system. Using Tsebelis' (1990) metaphor: the EU has to be seen as a nested game. Actors are playing different games simultaneously, and these game-situations have in turn different consequences for each player in regard to payoffs. EU negotiations are a continuous process, since they are often affected by previous rounds.

In other words, in an iterated or repeated game time matters, because without commitment no rational player would keep agreements, unless s/he considers it to be in his/her current and future interest to do so. The endeavors of the past constitute a kind of a sunk cost and are treated as such (Binnmore 1991, 123).

THE EMPIRICAL ANALYSIS: BARGAINING GAMES AND SINGLE-PEAKED PREFERENCES

Empirical evidence is provided by analyzing two key negotiations in the fisheries policy field, which illustrate how integrative and distributive bargaining work in practice. The first key fisheries negotiation deals, in EU-jargon, with the settlement of the common organization of the market and the structural policy. It constitutes an integrative bargaining situation, in which all negotiating parties can benefit from an agreement and consequently prefer it to the status quo. This translates simply as: some member states should get financial aid for modernizing their fishing fleet (structural policy) but in return they would have to accept a common market for fish products, which met the preferences of other group of member states. Since these issues were linked, from the perspective of the six parties involved, an agreement represented a Pareto-improvement.

The second negotiation situation on the settlement of the conservation and management policy deals with a distributive bargaining situation. Also in this bargaining situation, all players have a common interest in coordinating their strategies, since an agreement will provide all players with a better payoff. However, they disagree, they have different preferences concerning how to coordinate. Distributive politics is about sharing benefits and burdens. Many of the crucial issues of interdependence, and likewise of EU negotiations, focus on the oldest question of politics: "Who gets what, when, how?" (Lasswell 1971). Distributive bargaining is controversial by nature, because the concern of each member state is to maximize its own benefits and minimize its own losses. Most negotiations in the EU represent an integrative or a distributive bargaining game. The CFP, as many other policy issues in the EU, deals with one of the basic political subjects, namely the allocation of a value among member states. When the member states negotiate on the division of the structural funds, or on the national distribution of the EU budget (payers vs. receivers), a similar situation of distributing a value among themselves can be found. Therefore it does not matter which policy field is analyzed, as it is expected that the findings can be generalized for other EU negotiations about policies incurring expenditure.[12] Whereas the integrative bargaining situation involves internal issue linkage, understood as bargains within the same policy area providing a give-and-take-basis for the trade of concessions, in the distributive bargaining situation the focus is on division and on the bargaining power of single member states.

Estimating the policy positions of players allows a systematic description of preferences of member states and their relationship. Systematic descriptions of an actor's preference are based on the notion that preferences are single-peaked, that is an actor has an ideal or most preferred point, which can be depicted as a point on a line, with regard to a particular issue, and that other policy options can be systematically compared to this ideal point in terms of their closeness to it. The concept of the policy distance between at least two policy options is an entirely theoretical construct used in spatial models of political competition, which allows the different sets of actors' preferences to be illustrated in a graphical form.[13] The capacity to locate actors' policy positions, that is the ordinal ranking of actors according to their preference intensity (high or low), is a means to assess the extent to which preferences differ. This pattern varies on different issues. Such plots can be used to explain how coalitions between member states are formed when different issues are linked, an example of this will be given in chapter 3.

In *figure 3* the intensity of preferences of various member states has been plotted for two different cases, namely the structural policy issue and the common market organization. The preference intensity (high *vs.* low) is depicted along the vertical axis. Low values in the scale refer to a low preference on that issue, and high intensity indicates high preference salience for member states ($MS_1, \ldots, MS_6$), which are situated along the horizontal axis at the position that represents each member state's ideal point.

In the first example (A), the intensity of the preference for the structural policy issue is very high for MS_5 and MS_6, but low for the remaining M_1 to MS_4. In contrast, in the case of common market organization (B) the MS_1 to MS_4 have a very high preference, whereas the preferences for MS_5 and MS_6 arc low.

Figure 4 illustrates graphically the configuration of member states' preferences in the distributive bargaining game. The intensity of the preferences concerning the application of the principle of equal access and the division of total allowable catches is low for a majority (MS_1 to MS_6), but high for a minority (MS_7, MS_8, MS_9). In contrast to the first fisheries negotiations, this second bargaining game deals with only one-policy issue. As a result, it is expected that the negotiating tools or bargaining strategies used in integrative and distributive bargaining situations may differ from each other.

Estimating the policy positions of member states comprises two distinct analytical steps: the first step is to locate the preferences of member states in a policy space. The next step involves an interpretation of the substantive meaning of the policy positions space, which is dealt with in the two next chapters. Chapters 3 and 4 deal with actors' preferences and the institutional

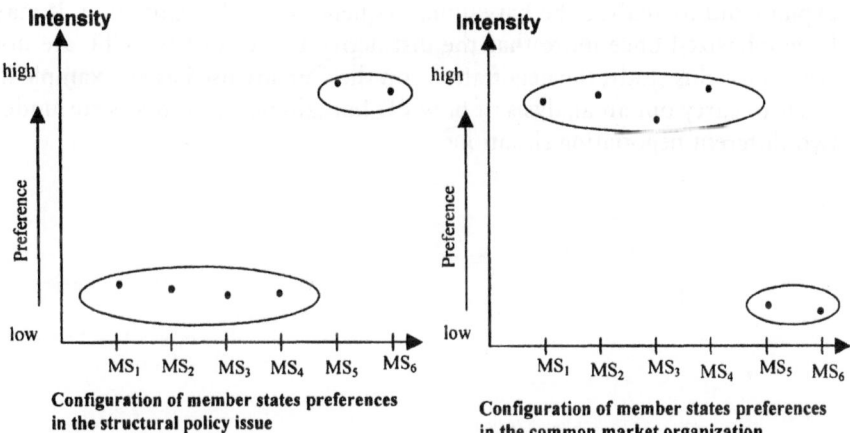

Figure 3: Configuration of member states' preferences in the integrative bargaining game

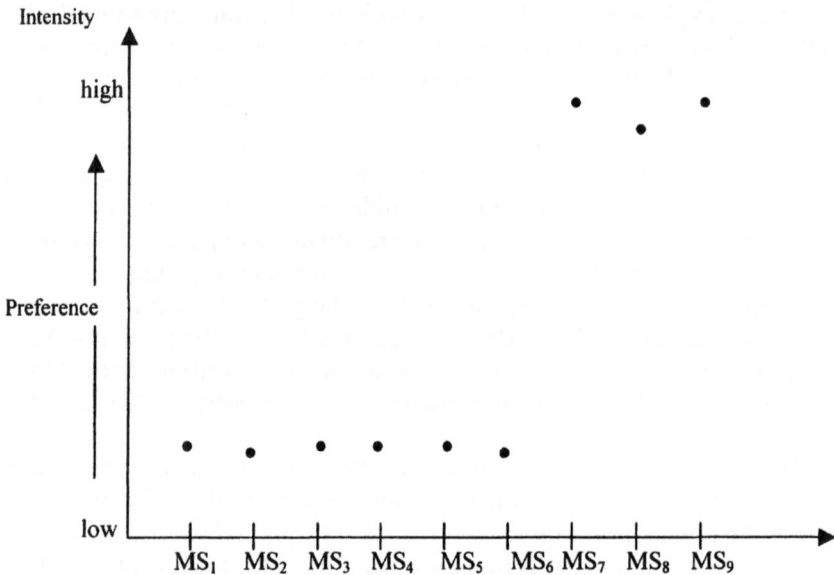

Figure 4: Configuration of member states' preferences in the
distributive bargaining game

setting in integrative and distributive bargaining games. In both negotiations
unanimity voting is assumed to be the constant parameter, only the number
of players is changed in order to test the difference it makes. The key question explored in these two fisheries negotiations is whether the conceptual
framework for analyzing negotiations in the EU developed earlier can help
the understanding of the actual policy configuration. Its objective is to
explain and to analyze the bargaining sequence of each negotiation. It may
be emphasized once more that the distinctive features of the CFP are not
central for this study, the negotiations on the CFP are used as an example in
order to carry out an analysis of how EU bargaining outcomes occur under
two different negotiating situations.

CHAPTER 3

Actors' Preferences and the Institutional Setting in Action
An Integrative Bargaining Game

NEGOTIATIONS IN THE EU AS A COOPERATIVE GAME IN PRACTICE

There are two general types of games in game theory: cooperative and non-cooperative. Whereas in cooperative games the players are able to make contractual agreements that are binding, in a non-cooperative game the players make their choices independently and cannot make binding agreements. The negotiation process in the EU can be looked upon as a cooperative game. Actors coordinate their strategies through binding agreements before and during the negotiations, and act jointly to maximize their preferences. Communication between the actors is allowed during this process. The core of the game is the decision to cooperate because the relative benefits of cooperation are assumed to be higher than the costs of non-cooperation.

In the EU negotiation process, national government representatives place their policy positions at the bargaining table. To reach a bargaining outcome, the negotiation process has now to center on compromising between the different policy positions. These policy positions define the bargaining space available for coming to an agreement that produces advantages for all involved parties. In any negotiation process, two central questions have to be considered: first, what deals are feasible and how an optimal choice can be made from the feasible set. The following integrative bargaining game will try to demonstrate how outcomes take place in the EU when Pareto-improvements are expected as a result of the coordination of policies. Thereby the process of finding an acceptable compromise formula for all involved players and the negotiating tools used will be a central issue in the following analysis, in order to find out which variables influence bargaining outcomes in the EU.

A meeting of the Council usually starts with a *tour de table*, where each minister states his/her position, and often s/he only repeats what has already been said by his/her civil servants at meetings of the working group or of the

COREPER. At the working groups level, however, member states do not try to reach an agreement where sensitive issues are concerned. They prefer to reserve their positions until the meeting of ministers. There are two different reasons for this in the fisheries policy. First, the national fishing industries might interpret concessions at the working groups level as a sign of political weakness. Second, the strategy behind reservations until the meeting of the ministers is the hope that each member state will be able to make trade-offs with other member states (Holden 1996, 212–213). In the case that an agreement cannot be reached, the presidency of the Council adjourns the negotiations and arranges for restricted sessions (bilateral meetings) in order to find a compromise. A bilateral meeting of 15–20 minutes, with each member state trying to set out the bargaining range, means a total of about five hours or more, given that there are now thirteen member states interested in fisheries, since Luxembourg and Austria have no stakes in fisheries. This explains partly why Council meetings take so long.

The next stage in negotiations is an attempt to find a final compromise benefiting each player. Thereby, the parties involved may spend several meetings drafting the final regulation. In practice, national government representatives often spend several months or even years in bargaining with each other. In the case of fisheries, the ministers and representative from the Commission may spend 28 hours or more in what is known as "marathon meetings"[1] bargaining, for example on the allocation of fish quotas. This practice can be explained by the fact that each minister frequently emphasizes that s/he does not agree with the proposal, but will be ready to accept it only at great sacrifice and only if all other ministers accept it as it stands. Marathon negotiations are often the penultimate bargaining session of the Council. When meeting at the Council, ministers are frequently trapped by a commitment to continue the negotiation to avoid adjourning the meeting. The whole atmosphere is directed towards resolving the differences and finding formulae of agreement, even if this requires working through the night or prolonging the meeting for several days. The invention of marathon meetings was based on the expectation that the sheer length of a meeting would force the ministers to come to a decision and that when they were tired enough, they would change their negotiating position, being conscious of the fact that the forest (EU) was more important than the single tree (member state) (Hayes-Renshaw and Wallace 1997, 61; Westlake 1997, 116). Furthermore, marathon negotiations dramatize for the domestic constituency just how hard a minister has fought to defend the national negotiating positions in a certain issue.

THE NEGOTIATIONS ON THE SETTLEMENT OF THE STRUCTURAL POLICY AND OF THE COMMON MARKET ORGANIZATION: MEMBER STATES' POLICY POSITIONS

The first negotiation on fisheries dealt with represents an integrative bargaining game. All the actors expected that a coordination of their structural policy for modernizing their fishing fleets and the settlement of a common market organization for fisheries products would give them an increase in gains.

The initial negotiating positions of member states in regard to structural policy and common market organization will be outlined to demonstrate how much a national delegation was willing to accept. In the first fisheries negotiation concerning the structural policy the actors can be differentiated in respect to their policy positions as minimalists, Germany (MS_1), the Netherlands (MS_2), Belgium (MS_3), and Luxembourg (MS_4) and maximalists, France (MS_5) and Italy (MS_6). Although Luxembourg has no stake in fisheries, it had a clear position on the structural policy due to its position as a net contributor to the budget of the EU. In the issue of the common market organization, however, constellation of preferences was totally different. Here MS_5 and MS_6 represented the minimalist actors preferring less liberalization, whereas the other four member states, MS_1, MS_2, MS_3 and MS_4 wished to have more liberalization of fish markets.

Table 2 formalizes the transitive preferences of member states on the structural policy and on the common market organization. In a negotiating situation all the actors (MS_1, MS_2, MS_n) have different preferences for the bargaining outcome and rank the alternatives. It is assumed that there are three alternatives, $\{x,y,z\}$ in the structural policy and $\{m,l,s\}$ in the common market organization, to which all member states had complete and transitive preferences.

MS_5 and MS_6 can make statements like "I prefer x to y" and "I prefer y to z". Here the financing of the structural policy by the Community is repre-

Table 2: Member states' transitive preferences on the structural policy and on the common market organization

MS	Issue 1 (Structural policy)	ISSUE2 (Common Market Organization)
MS_1 (D)	z>y>x	m>l>s
MS_2 (NL)	z>y>x	m>l>s
MS_3 (B)	z>y>x	m>l>s
MS_4 (L)	z>y>x	m>l>s
MS_5 (F)	x>y>z	s>l>m
MS_6 (I)	x>y>z	s>l>m

Legend of the table:

	MS	Member State
	x	100 percent Community's financing
	y	50 percent
	z	0 percent
	m	more market liberalisation
	l	less market liberalisation
	s	Status quo

sented as follows: $x = 100$ percent, $y = 50$ percent, and $z = 0$ percent or maintenance of the status quo. Thus for MS_5 and MS_6, x is better than y and y better than z $(x>y>z)$. In contrast, the other four actors had a completely different preference on this issue. MS_1 (D), MS_2 (NL), MS_3 (B), and MS_4 (L) preferred z to y and y to x $(z>y>x)$.

Concerning the common organization of the market the negotiating positions of the actors differed completely from the structural policy issue. MS_1, MS_2, MS_3, MS_4 preferred m (being more market liberalization) to l (less market liberalization) and l to s (status quo), MS_5 and MS_6 preferred s to l and l to m ($s>l>m$). *Table 2* shows that the actors opposing a structural policy in fisheries (Issue$_1$), wished to have a common market organization (Issue$_2$), and those players in favor of a structural policy vetoed the common market organization. Taking these policy positions as a starting position for the negotiation process, the question arises which actors were able to push through their preferences, and how the preferences of the Commission and the institutional setting interacted with the preferences of member states.

THE PREFERENCE OF THE EUROPEAN COMMISSION: AN AGENDA-SETTER USING THE WINDOW OF OPPORTUNITY

The Commission prefers to strengthen the European integration process, which allows it to expand its power by allocating competencies to the supranational level. The Commission is more than a simple agent fulfilling the tasks delegated by the principals. The Commission is modeled as a purposeful opportunist trying to expand its power. It can make use of its agenda-setting power and act as a "policy entrepreneur"[2] (Kingdon 1984) through innovative policy proposals that loosen the gridlock between member states and promote a more substantial agenda. This implies that the Commission exploits the original demands of single member states by issuing proposals that anticipate rationally a majority among the integrationist actors. Lindberg and Scheingold (1970, 93–94) pointed out that if the Commission is to play an active rather than a passive role, it must make creative use of its own resources for influencing the bargaining behavior of the member states. This can be the clear articulation of a goal, the building of coalitions, or the proposal of package deals. The question arises which of these strategies did the Commission engage in the fisheries policy.

The Commission was able to use the window of opportunity and its agenda-setter power in the fisheries policy field. It combined the demand of one group of actors for a structural policy and that of the other group for a common market organization into a package and added in 1968 a proposal for the basic principles for a common policy. The latter was published as *"Report on the Situation in the Fisheries Sector of EEC Member states and the Basic Principles for a Common Policy"*. This proposal introduced the principle of equal access to fishing grounds. The Commission used this window of opportunity by acting as a political entrepreneur looking for new competencies. The Commission realizing the uniqueness of such a policy window, exploited it to its own advantage, and proposed a broader change in policy. Since the member states offered no resistance to the original proposal of the Commission on a common fisheries policy, it was a topic of the discussion for the settlement of the conservation policy in the following bargaining round. This case illustrates that member states can instruct the Commission to prepare a text, but they have no control over its content.

This apparent indifference of member states to a complete transfer of this policy field to the supranational level can be explained by the fact that member states expected to get mutual gains from coordinated action in the fisheries policy. Since some countries wished to acquire statutory rights of access for their vessels to Norwegian, Greenlandish, British, and Irish waters after the introduction of the 200 mile EEZ, a common fisheries policy was considered to be better than unilateral actions to cope with changes in the international fisheries regime. Furthermore, most attention was centered upon the compromise on structural policy and the common market organization. The discussion on how to give equal access to fisheries resources was relegated to the next negotiation. National government representatives, who are concerned essentially with short time horizons and with the next elections, paid little attention to the long time perspective of the issue of equal access to fisheries. In contrast, the Commission, being a bureaucracy with a long time horizon, took advantage of this window of opportunity to expand its competencies and to maximize its powers.

THE BARGAINING PROCESS: ACTORS' PREFERENCES AND THE INSTITUTIONAL SETTING IN ACTION

At the beginning of the bargaining game, there was the following policy positions of member states: France and Italy were prepared to accept trade liberalization and harmonization on price support and market organization, if, in compensation, agreement was reached on a structural policy for the fishing industry, to be financed by the Community. Both countries wished to obtain financial support for their fishing industry to assist them in coping with the increasing competition for fish as anticipated by the common market organization. On the other hand, Germany, the Netherlands, and Belgium had well-organized industries and expected to profit from trade liberalization. They did not want more interventionist policies that would increase the Community's budget and automatically their own contributions, while aiding their weaker competitors in modernizing their fishing fleets.

Having these bargaining positions how could a compromise formula be agreed upon and how could the institutional setting influence the actors' negotiating positions in this first negotiation on fisheries policy? It is generally assumed that as every actor has a veto power, the use of the formal rule of unanimity voting can lead to an impasse in negotiations. Due to the fact that the preferences of member states are strongly polarized in the issue under discussion (one group of actors is in favor of a structural policy for fish products and against a common market organization, while the other group is in favor of a common market organization, but against a structural policy), and the number of actors was relatively small, the parties involved used the building of coalitions as a negotiating tool. Thus two coalitions emerged. *Coalition A*-France and Italy, and *Coalition B*-Germany, the Netherlands, Belgium, and Luxembourg. These two coalitions were formed to increase the bargaining power of the actors whose interests were similar.[3]

The primary objective of *Coalition A* was to have a structural policy at the supranational level, in order to obtain Communitarian aid for the modernization of their salt cod and tuna fleets, to improve shore based infrastructures and, if necessary, to redeploy redundant labor force. France and Italy were ready to accept a common market organization if minimum prices and intervention arrangements (with compensation for products withdr. wn from the market at minimum prices) were introduced. Initially, Coalition A wished Community funds to cover the complete cost of modernization. The position of *Coalition B* was to avoid excessive Communitarian intervention or protection and to get a minimal coordination of the structural policy. This was clearly reinforced by the fear that *Coalition B* would have to pay the largest share of the new system with a higher contribution to the EC-budget (Leigh 1983, 28; Lequesne 2001, 35).[4] *Coalition B* had no interest in adopting a structural policy benefiting primarily *Coalition A*. This is also in line with the König and Bräuninger (2001, 12) tax payer model, in which they assert that the fewer advantages a member state has from a certain policy, the less willing it will be to contribute financially to that policy. Since *Coalition B*, especially Germany and the Netherlands, had a modern and well-organized fishing industry, it required little financial assistance from the Community. Consequently, *Coalition B* had a clear preference for less rather than more financing of the structural policy.

The two coalitions and their preferences on the structural policy are described in *table 3*. There are two factions, one for more and one for less financing. The *"less financing"* faction is made up of net contributors to the EC-budget, *Coalition B* and includes the countries that on the whole prefer less Communitarian financing as they know that they will have to pay the bill. Their best alternative would be 0 percent Community financing, second best 50 percent and worst 100 percent Community financing.

The *"more financing"* faction is made up of *Coalition A* constituted by member states which generally take a more protectionist position in different policy issues and favor budget increases. The best alternative for this faction would be to have a 100 percent Community financing for their fishing industry, next best is 50 percent and worst is keeping the status quo (0 percent).

Table 3: Ranking of alternatives by coalitions on the structural policy

	FACTION	
Ranking	*Coalition B* less financing	*Coalition A* more financing
Best	0%	100%
Middle	50%	50%
Worst	100%	0%

A SHORT ADDENDUM ON COALITION-BUILDING

A coalition is set up by a binding arrangement among players. This kind of arrangement is possible when involved parties expect to obtain joint gains and form a concerted action. Coalition building is an instrument used in negotiations for the achievement of mutually beneficial gains. Actors aggregate their preferences and form a coalition to force their decision upon others, when they wish either to reach or to prevent an agreement (Gamson 1961, 374; Dupont 1996, 49). A coalition incorporates the preferences of each single coalition member, as this is easier to negotiate than having to consider the sum of individual preferences. Actors expect to maximize their total payoff when they decide to coordinate their strategies.

The expectation that minimal winning coalitions[5] would emerge is associated with the work of Riker (1962, 32–33), who stated that coalitions are as large as participants who believe that they will ensure winning. The formation of coalition governments has been the object of theoretical and empirical analyses. These studies explain that one motivation for forming a coalition is office-seeking, that is the primary aim of politicians is to get into office (Gamson 1961; Riker 1962; Laver and Schofield 1990). The second motivation to build a coalition with certain partners is governed by parties' preferences over alternative policies. According to Plott (1967) actors evaluate alternative coalitions depending to a certain extent on their preferences on the alternative policy proposals advocated by the different coalition members. Under majority rule within various specific frameworks, policy bargaining between different political parties rarely leads to stable coalition structures (Schofield 1983).

The few studies available on this subject have shown that coalition building in the EU follows lines similar to those outlined above. It has been stated that coalition-building in the EU is based primarily on certain criteria, such as geography (Northerners versus Southerners), size of the countries (for example Benelux small, Germany, France, and Great Britain large), and on budgetary issues (net contributors versus net receivers) (Hosli 1996; Spence 1995). It may be noted that coalitions do not remain stable, but change according to the issues under discussion. In the bargaining game analyzed in this chapter, the budgetary issue and the transfer of financial resources (subsidies for the modernization of the fishing fleets) from some member states to others were the main incentive for building coalitions.

OVERCOMING THE UNANIMITY TRAP: THE USE OF ISSUE LINKAGE

In an issue linkage situation an actor gives up something of value in one issue if in exchange s/he receives concessions in another, for him/her, more important issue. Decisions are taken on different issues at the same time so that they can be accepted by all negotiating parties. Without the existence of issue linkages, member states profiting from one policy issue, but getting negative payoffs from another, would obviously only support the first issue (Hosli 1996, 256; Milner 1997, 109). Numerous studies on issue linkage in international relations indicates that issue linkage might be successful in a bargaining situation when the parties involved value the issues differently (Conconi and Perroni 2001; Lacy and Niou 1998; Keohane and Nye 1977; Tollison and Willett 1979; Weber and Wiesmeth 1991). Sebenius (1983) demonstrated how adding issues into a negotiation can yield joint gains and change the zone of possible agreement among the parties involved. In public choice literature, log-rolling is the term used to explain this situation. In a log-rolling bargaining situation a trade between actors (A and B) takes place. Both actors agree to vote against their preferences, because they will receive something in return. Log-rolling can be made within a single policy area or over a wide range of areas. Tullock (1998, 104) distinguished between explicit and implicit log-rolling. In the first form one actor (A) agrees to vote against an issue s/he actually favors, in return for the concession of the other actor (B) to vote at some later point for an issue actor A prefers. Two items are combined in the implicit log-rolling bargaining situation, which was an important tool for reaching agreement in the fisheries policy. Whereas in an implicit log-rolling situation, two items are stuck together into the same bill.

In the EU, issue linkage has been often applied to reach agreement. The first important example of package deal or issue linkage dates back to 1955 and concerns the Euratom and the EEC. Similar to the bargaining game analyzed here, there were two factions. Germany with its strong export orientation was in favor of a EEC, but was less interested in the Euratom proposal. While France feared possible negative effects of a common market, but wished to develop its nuclear energy program (cf. Weber and Wiesmeth 1991, 256). Two forms of issue linkage, external and internal linkage, have been used in the fisheries. Using external issue linkage, a member state may link a policy sector not connected with fisheries to obtain a decision on a fisheries issue, whereas in internal linkage two different issues within the fisheries policy are linked. In the fisheries negotiation on the settlement of the structural policy and on the common market organization, due to the highly polarized preferences, or the strong preference relation in two different issues, internal issue linkage was used as a negotiating tool for overcoming the unanimity trap. Detailed analyses of other policies in the EU revealed that issue linkage has often been a very useful bargaining tool.[6]

Issue linkage, however, is not always possible and also becomes a complex and politically risky strategy when the domestic losses and gains produced by linkage are valued differently by the domestic constituencies.

EXPLAINING THE BARGAINING OUTCOME: COALITION BUILDING
AND ISSUE LINKAGE

The bargaining outcome of the first fisheries package introduced the two first pillars of the CFP: a common market organization for fish products and a structural policy. Although agreement had to be reached with unanimity voting, the negotiations were quite short and easy since the number of parties involved was relatively small and it was possible to form two coalitions. *Coalition A* consisted of France and Italy, and *Coalition B* of Germany, the Netherlands, Belgium, and Luxembourg. The two coalitions disagreed on the scope and the nature of the structural policy and the common market organization. The maximalists on the structural policy (*Coalition A*) pushed for a 100 percent Community's financing of structural policy of which they were the direct beneficiaries. On the other hand, the minimalists (*Coalition B*) wished to minimize the costs of a common structural policy for the fishing industry, and in the case of the common market organization, there was an inverse situation.

The Commission preferred agreement to no agreement. Its primary preference was to expand its power, and for this purpose developed a broad proposal on fisheries, in which the basic principles for a common policy were projected. Under unanimity voting when coalitions are built, the role of the Commission consists essentially of mediation and of avoiding a stalemate. Through issue linkages, the Commission was able to prepare a compromise formula, which could be accepted by all players. Thus, the common market organization for *Coalition B* was only accepted by *Coalition A* through the settlement of a structural policy for financing the modernization of the fisheries sector in these countries.

In order to reach a compromise between the two parties, a typical EU result came out: in the structural policy fifty percent of the financial fund necessary for modernizing the fishing fleets would come from the Community's budget and fifty percent from the member states concerned. In the common market organization liberalization for fish products was introduced, but under the condition that withdrawal prices, that is the minimum price below which fish can not be sold, for major fish species were established. As with agricultural products, when the market price fell producers' organizations could take fish from market. This constituted a compromise solution between *Coalition B* favoring an interventionist policy and *Coalition B* wishing to avoid a further increase in the EC-budget. The members of *Coalition B* were all net-payers to the EC's budget. Thus a structural policy financed totally by the EC would have automatically increased their contribution. This illustrates that actors with net gains from a European policy will favor transfer to the supranational level, whereas actors being net contributors to the EU-budget are rather reluctant supporters of any new initiative inter-linked with new financial costs.

By early 1970, an agreement on this compromise solution could be reached, because the EC had to give the Commission a mandate of negotiations with GATT concerning a liberalization of trade on fish products. Furthermore, delaying agreement until the beginning of accession negotiations would have meant that accession countries (Great Britain, Ireland, Denmark, and Norway[7]) would have demanded that they be consulted, and a package deal taking into account the views of the accession candidates would have been far less favorable to the original six member states. The shadow of the future gave a decisive impetus to reach agreement. This external policy input led the *Coalition B* to accept a compromise, since at that time, the accession candidates had not only a strong interest in fishing issues, but also rich fishing grounds. On June 30, 1970, the day before accession negotiations started, the Six reached an agreement on this issue. Even though the compromise was less ambitious than the proposal of the Commission, it represented a consensus and created a *fait accompli*, which had to be accepted by the accession candidates. It may be noted that the way in which the new regulations were hastily put together, created misgivings in the candidate countries and gave rise to acrimonious debate on the fisheries issue after the accession, affecting future negotiations.

The first CFP bargaining outcome can be seen as a classic EU compromise achieving something for each national government and which can be accepted by the domestic public opinion and constituencies as a gain, namely liberalization of fish markets for some, and financial payments for the others. For *Coalition A* the policy agreed upon was better than the status quo. *Coalition B* accepted the demands of subsidizing the modernization of *Coalition A*, because in return it would obtain a trade liberalization of fish products. It was a typical situation of EU bargaining: member states exchanged concessions in one issue-area where their preferences were relatively weak for concessions in another area where the salience of an issue was higher. Or as Moravcsik (1993, 505) points out that in case a set of arrangements is taken separately, it would not be accepted by at least one actor, but they may create gains for all when accepted as a package deal or as a positive sum issue linkage, that is mutually advantageous for all parties involved.

CHAPTER 4

Actors' Preferences and the Institutional Setting in Action

A Distributive Bargaining Game

THE SETTLEMENT OF THE CONSERVATION AND MANAGEMENT POLICY: THE
ACRIMONIOUS NEGOTIATIONS ON HOW TO DIVIDE THE FISHERIES RESOURCES

After the settlement of the structural policy and of the common market organization, in 1976 it was easy for the member states to come to an agreement on an external fisheries policy. These agreements with third countries were of advantage to all those member states with a long distance fishing fleet. In contrast, the negotiations on the conservation and management policy initiated in December 1976 took almost seven years until an agreement could finally be reached in January 1983.

Any negotiation can be dissected into a two-step process. Whereas in the first step the focus is on identifying and defining the parameters of disagreement and the set of possible outcomes configurations over which involved parties can agree upon; in the second negotiating stage, actors have to engage in bargaining in the traditional sense of dividing the gains from agreement among themselves (Cross 1996, 156). Although players recognize the potential for mutual gain in coordinating their policies, they still have to agree on its distribution before the gain can be realized. In the bargaining situation analyzed in this chapter the central issue, when settling the conservation and management system, was to ensure that fisheries resources were not depleted in the present at the expense of the future. The recognition of the necessity for conservation and management system of fisheries resources, however, was just the starting point of the negotiations. The member states still had to agree on how to fix the national quotas and on how to settle the rules on access to fishing grounds. The negotiations were acrimonious, because they dealt with a distributive bargaining situation, that is to say, on how to distribute fisheries resources among member states.

There are two distinct negotiating phases in this distributive bargaining game: in the first phase, Great Britain and Ireland opposed any agreement, and in the second Denmark disagreed with the compromise formula found

65

for the distribution of fish resources and also with the settlement of fish boxes. During the negotiations, the role played by veto players was fundamental. In this chapter the theory of two-level games will be used, since it is a helpful way to explain why a player employs his/her veto power, and may also provide a plausible explanation of how domestic politics influence an actor's negotiating position at the supranational level. The question is whether the negotiating tools used to reach an agreement in this distributive bargaining were different from those in the integrative bargaining game and how deadlock situations in the EU may be broken down when member states veto agreements.

MEMBER STATES' NEGOTIATING POSITIONS: RESTRICTED AND UNRESTRICTED APPLICATION OF THE EQUAL ACCESS PRINCIPLE

At the opening bid, the negotiating position of the member states has to be outlined, as this is what determines the final outcome. It is assumed that all member states wish to have a conservation and resource management policy in fisheries in order to avoid the total depletion of fish stocks. They have different preferences, however, on how to reach this goal. Actors can express their preferences over the set of outcomes and can rank alternatives from the most to the less desired outcome. *Table 4* formalizes the transitive preferences of member states on the application of the equal access principle. In this bargaining situation, it is assumed that three different alternatives $\{r,u,s\}$ are possible. The table shows that Germany (MS_1), the Netherlands (MS_2), Belgium (MS_3) Luxembourg (MS_4), France (MS_5), Italy (MS_6), and Denmark (MS_9) preferred u (unrestricted application of the equal access principle) to r (restricted application of the equal access principle)and r to s (status quo). Whereas Great Britain (MS_7) and Ireland (MS_8) preferred r to u and r to s ($r>u>s$).

The two veto players, Ireland and Great Britain, caused a fundamental split between two groups of the member states during the first negotiating phase. The majority of the member states claimed that all the resources in the 200 mile EEZ should be managed by quota allocations at the supranational level. They were in favor of an unrestricted application of the equal access principle. The minority, Great Britain and Ireland, demanded that the application of the equal access principle should be restricted in order to reserve certain areas for their exclusive use. In addition, they demanded a particularly high share of the TACs. The initial reluctance of both countries to a conservation and management policy that would provide equal access to fisheries resources, was because their coastal waters are rich in resources, and therefore there was no necessity for their fishing vessels to enter the waters of other member states. Belgium, the Netherlands, France, Germany and Denmark, however, had narrow coastlines or poor fish stocks, and were unable to satisfy their fish needs in their own coastal waters.

Table 4: Member states' transitive preferences on the application of the equal access principle

MS	Preferences on the Application of the Equal Access Principle
MS$_1$ (D)	u>r>s
MS$_2$ (NL)	u>r>s
MS$_3$ (B)	u>r>s
MS$_4$ (L)	u>r>s
MS$_5$ (F)	u>r>s
MS$_6$ (I)	u>r>s
MS$_7$ (GB)	r>u>s
MS$_8$ (IRL)	r>u>s
MS$_9$ (DK)	u>r>s

Legend of the table:
MS member state
r restricted application of the equal access principle
u unrestricted application of the equal access principle
s status quo

Before the settlement of the 200 mile EEZ most of the catch of the Community had been fished in British and Irish waters, and both these countries were going to lose the most from a common management of fish stocks in their waters. For this reason, they had a very strong negotiating position at the bargaining table. This situation produced an asymmetrical pattern of fishing interests and explains why the majority of actors (MS$_1$, MS$_2$, MS$_3$, MS$_4$, MS$_5$, MS$_6$, and MS$_9$) was in favor of the unrestricted application of the principle of equal access, while two actors (MS$_7$ and MS$_8$) preferred a restricted application of it.

THE PREFERENCE OF THE EUROPEAN COMMISSION: THE IMPORTANCE OF HAVING A QUOTA SYSTEM

The proposals of the Commission on the conservation policy concentrated on sharing the available fish stocks by a total allowable catches (TACs) system. The choice for the catch quota system has often been criticized as out of place, because it is difficult to implement these quotas through appropriate controls. It has also been emphasized that restrictions on the number of vessels permitted to engage in a fishery through a licensing system are easier to enforce than TACs. The question arises why was it so important for the Commission to have a management system based on TACs and national quotas.

The Commission adopted an approach laid down in the Law of the Sea Convention, whose article 61 states that the coastal state shall determine the TAC of the living resources in its EEZ. The Commission proposed that TACs should be established for the EC's principal stocks on the basis of scientific advice from the International Council for the Exploration of the Sea and from the Commission's Scientific and Technical Committee. The most important feature of this management system was that the Commission itself would monitor catch levels and decide when it should call upon national authorities to close a fishery when the TAC had been reached. Furthermore, as the TACs system was seen as insufficient for a proper conservation of EC's fish stocks, the settlement of national quotas would make it easier to keep catches within the TACs (Leigh 1983, 89). Thus TACs and national quotas together would constitute the core of the EC's conservation and management policy supplemented by other technical measures such as minimum mesh and fish sizes, closed seasons and areas.

In this way, the proposal of the Commission on the settlement of a TACs regime and national quotas was motivated by utility maximizing considerations. Since a quota system is defined and enforced centrally only the Commission could propose quotas for the entire EC. The information required for the operation of the quota system is communicated to the administration in Brussels to be monitored there. The acceptance of the catch quotas system would automatically imply a common management of fisheries resources. For this reason a centrally monitored quota system allowed the Commission to push through its preferences, that is to expand its power and strengthen the European integration process. Moreover, it allowed the Commission to increase its informational asymmetry vis-à-vis the member states and to make full use of its formal agenda-setting power. Tullock (1981, 190) argues that when the agenda-setter has a complete knowledge of everyone else's preferences, and all actors always vote for their preferences, the agenda controller is the only strategist among the actors' group. This illustrates the decisive influence that an agenda-setter, such as the Commission, can have in a bargaining process.

Nevertheless, member states bargained from 1976 and were unable to reach agreement on sharing quotas in EC waters between member states, because the apparently technical rubric TACs and quotas disguised a political problem of resource distribution between member states with an interest in fisheries issues.

BARGAINING RANGE AND THE ZONE OF AGREEMENT IN A DISTRIBUTIVE BARGAINING GAME

In any negotiation process, actors initially ask for more than they expect to obtain, because they know how far they are willing to go in terms of concessions. Each actor has a reservation point, a reservation level expressing the minimal deal that s/he would accept. The crucial problem in a bargaining process is to identify a zone of possible agreement, the so-called contract zone or bargaining range. This zone of possible agreement, where for each actor a set of possible agreements is superior to the non-cooperative alternatives to an agreement, includes all the possible bargains that both sides prefer to the status quo (Sebenius 1991, 212–214; Jönsson 2001, 2).

In a distributive bargaining game, players act both cooperatively and competitively. They cooperate in the sense that they try to increase jointly the gains from an agreement, but at the same time they act competitively since each player tries to obtain the largest share of a resource (for example budget, structural funds, or national quotas) so that his/her preferred outcome is reached. In the EU bargaining process the object for each party is to get a maximum outcome for their own preferences. To achieve this aim they have to be able to convince their counterparts that they will not hesitate to use their veto power in order to impose their interests. The hallmark of bargaining is the latent presence of coercion. Hence, the use of hidden or open threats and warnings are the central elements available to all actors in bargaining situations (cf. Eriksen 1999, 17).

NEGOTIATIONS WITH VETO PLAYERS: AN EXTENSIVE-FORM GAME

An explicit game theoretic formulation of the bargaining process may be helpful for a better understanding of the negotiation process in the EU, since game theory can describe negotiations in a somewhat more formal manner. The negotiation process will be modeled as a bargaining game of alternating offers and counteroffers between two actors, namely between the European Community (EC) and the respective veto player. The Rubinstein model, which has been applied successfully to international relations (Iida 1993, Mo 1994) and to domestic politics (Baron and Ferejohn 1989), will be used to illustrate this. In this model, two players at negotiations at the international level alternate in making proposals which the other party may accept or refuse. Each round of bargaining consists of two moves, offer and response. Perfect equilibrium outcomes lead to a bargain being reached which is an approximation to the Nash Bargaining solution[1], provided that the involved parties discount time at the same rate and also that the interval between successive compromise proposals is sufficiently small.

The sequence of moves in the negotiating process with a veto player can be illustrated using an extensive-form game. Even very simple bargaining games, however, can very quickly become quite complicated when represented in extensive form. Thus games are also often analyzed in a reduced form, the normal form, which represents a condensed version of a game and focuses on the choice of overall strategies and the payoffs associated with such strategies. In this chapter, extensive-form games will be used, since they allow one to identify how strongly one player prefers one outcome to another, when comparing two possible outcomes in a certain negotiation.

A two-person game in extensive-form is characterized by a game tree consisting of nodes and branches, a division of the nodes over the players and the probability of how distribution will be affected by each chance move. *Figures 6, 7 and 8* show three bargaining games, the first between the EC and Ireland, the second between the EC and Great Britain, and the third between the EC and Denmark. In the negotiations between the two first veto players it is assumed that both actors are monolithic, that is they have no internal divisions. In contrast, in the bargaining game with the third veto player, Denmark, internal divisions are included in the analysis of the bargaining process and this complicates and delays the negotiations.

Figure 5 shows that the bargaining game follows a sequence of moves which is a general feature of such negotiating situations. At the opening bid of negotiations, the veto player refuses to accept a change in the status quo (SQ), and therefore blocks an agreement. Once the veto player has chosen a position, s/he and the EC play a three-stage bargaining game. At the rhetoric stage, the veto player sends a message about his/her preferences to the EC. The EC, in turn, responds by proposing a compromise α_{nEC} at the proposal stage. At the approval stage, the veto player accepts or rejects α_{nEC}. If the proposal is rejected, the negotiations continue or the final outcome remains the status quo.

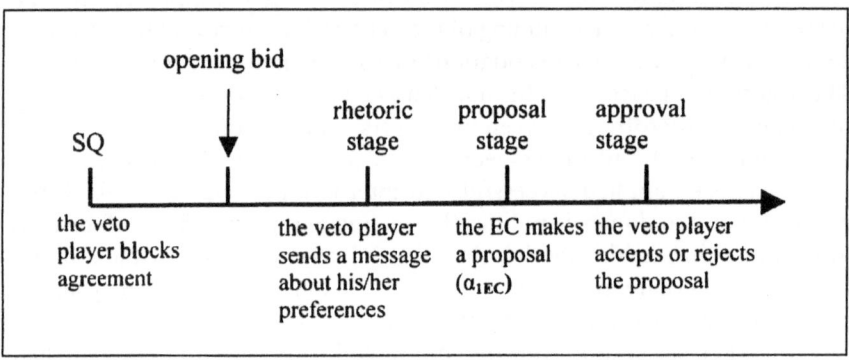

Figure 5: The sequence of moves in a bargaining game with a veto player

THE FIRST BARGAINING ROUND: GREAT BRITAIN AND IRELAND AS VETO PLAYERS

The initial negotiating position of the two veto players, Ireland and Great Britain, was to demand an EEZ of 50 miles within the future 200 mile EEZ of the EC. Furthermore, Great Britain wished to have the principle of juste return applied to fisheries. Since 60 percent of the Community catches was taken in the British 200 mile EEZ, Britain demanded 60 percent of Community allocations of fish stocks. This led to a hardening of positions. The other member states and the Commission refused to accept such a claim and

EC. In other bargaining situations, however, it is possible that the game will continue (t = 3) leading to new concessions until an agreement can be reached or to a deadlock situation, in the last case the status quo will be maintained and the negotiations broken down.

Obtaining the consent of this first veto player through side-payments had been relatively easy, but this did not terminate the negotiations for the EC, as the second veto player (Great Britain) had to be persuaded of the advantages of allowing access to its rich fishery resources.

The bargaining game between the European Community and Great Britain

When a member state finds itself in a minority position it can react in two ways: it can moderate its demands and move towards the other states or it may even harden its position. Britain chose the first option. After the legislative elections in May 1979, the new British government was ready to compromise on British claims. It no longer demanded an exclusive 50 mile EEZ, but only requested a dominant preference in the 12 to 50 mile band (Leigh 1983, 91–92). Moreover, the initial British demand for 60 percent of the Community's TACs was reduced to 45 percent. The EC, however, was only willing to offer an overall share of 31 percent of the Community's TACs.

Figure 7 is a formalization of the bargaining game between the EC and the veto player Great Britain. Once more, a game tree illustrates the different game situations. In the even-numbered period (t = 1) Player 1 (EC), who represents the negotiating position of the other eight member states, starts by making a proposal ($_{1EC}$), corresponding to 31 percent of the total allocation of TACs to Player 2 (Great Britain [GB in the figure]). Thereby, Player 2 has either the choice to accept or reject the proposal made by Player 1. If Player 2 accepts this first proposal, the endpoint of the negotiating game can easily be reached in a short period of time. However, this first proposal was rejected, so that Player 1 made a second proposal, namely (α_{2EC}). Each line projecting from the node represents an offer from Player 1 to Player 2. While Player 2 once more rejected the proposal, it took the initiative of demanding 45 percent of the total allocation of TACs. Player 1 rejected this demand and laid a third proposal on the negotiating table, the (α_{3EC}).

The final offer of the EC on the allocation of fish stocks gave Great Britain 36.1 percent of the total fish allocation, the highest quota of TACs granted to any member state.[5] In addition it obtained concessions in the form of a dominant preference for British fishermen in the Shetland and Orkney boxes. These boxes are fishing areas reserved for British fishermen.

At the opening bid, Great Britain had demanded 60 percent of the total fish allocation and an exclusive 50 miles zone. At the end of the negotiations it received only 36 percent of the total fish allocations in addition to the catches in the Shetland and Orkney boxes. Does this apparent switch in the British negotiating position mean that Player 1 was a strong negotiating

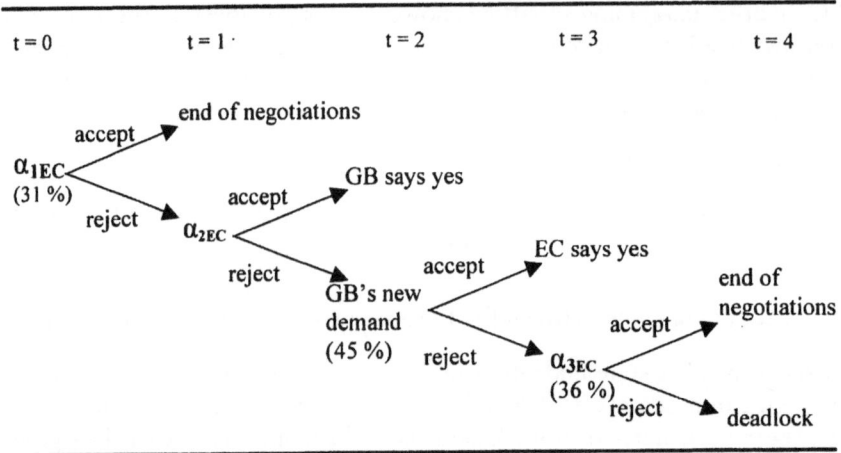

Figure 7: The extensive-form game between the European Community and Great Britain

player, who could easily push through his/her preferences, or is the final outcome of the bargaining between the EC and GB and rather an illustration of how a veto player push through his/her preferences in a more subtle way?

Player 2 (British negotiator) had realized that Player 1 (the EC) was not ready to comply to his/her demands for 60 percent of the total fish quotas. Player 2 then tried to push through his/her preferences more subtly by choosing the term dominant preference which for the British government was a way of reaching a reasonably satisfactory compromise that could be presented to the domestic constituencies. The term dominant preference translated into simple language means that restrictions of the access to British waters were settled by maintaining the Shetland and Orkney boxes as exclusive zones.

In order to induce Britain to accept only 31 percent of the total fish allocation and to agree to the conservation and management policy, the principle of equal access was weakened as a side-payment to Britain.[6] The readiness of Great Britain to cooperate can also be explained with another factor. During the negotiations, and after a long campaign, the British Prime Minister, Margaret Thatcher, obtained the agreement of other member states to an interim solution to the problem of British contribution to the EC's budget. At the beginning of the 1980s, due to the discussion on the British contribution to the EC's budget, the Community was literally paralyzed. Following a doctrine of juste return, Margaret Thatcher demanded her money back, and as she encountered some resistance from the other member states, she blocked the decision-making process. A Council resolution of 30 May 1981, tying a three-year-solution for the British budget, appeared to make a satisfactory progress in the formulation of a CFP. The linking of the fisheries policy to the British contribution to the budget of the EC led to more flexibility from the British government. As a part of a quid pro quo from Britain, the other member states expected that an agreement could be reached on the conservation policy, so that on January 1ˢᵗ,1981 a common overall fisheries policy could be settled (Leigh 1983, 84; Ritchie and Zito 1998, 160). Through this external issue linkage the deadlock situation could be broken once more.

When it seemed that the four-year deadlock could finally be broken, France rejected the compromise formula, refusing to negotiate over its historic rights of fishing within twelve miles of British baselines and the Council broke up in disarray. The reason for this short French blockade was the presidential electoral campaign. Shortly before a presidential election the French government representative was not willing to negotiate a compromise deal that the opposition could exploit for the electoral campaign. In December 1980, the electoral campaign for the presidential election had just started and the government felt vulnerable to criticisms of weakness from the Left. On the one hand, the Communist Party had demanded a French share of Community quotas of at least 20 percent, but in the Council discussions were converging on a French quota of less than 14 percent. On the other hand, the Socialist Party had a strong electorate in France's northern coastal regions, mainly in Brittany, where fishermen relied on the access to the waters of Cornwall, the Orkneys and Shetlands. Thus, one of the themes in the 1981 French Presidential election was the attitude of the French government towards Great Britain with President Giscard d'Estaing being accused by the socialist candidate, François Mitterrand, of a lack of firmness in negotiating with Great Britain over the CFP (Leigh 1983, 85; Shackleton 1983, 354). In the first half of 1982, with French Presidential elections over, the political climate was more propitious for a compromise.[7]

After this short French obduracy, there no longer seemed to be any obstacle for settling a conservation and management policy, since all the veto players had received side-payments in order to lift their vetoes. Nevertheless, the restrictions agreed to on the principal of equal access to fish resources in British waters, did not satisfy the Danish government, who rejected the whole compromise and insisted on the increase of the Danish quota of mackerel. Thus, agreement on the conservation and management policy could not be sealed, because of the Danish government's veto. This was the beginning of the last bargaining round.

THE BARGAINING GAME BETWEEN THE EUROPEAN COMMUNITY AND DENMARK

At the beginning of negotiations on conservation and management of fisheries resources, Denmark's negotiating position was in favor of unrestricted application of the equal access principle. Only after the veto of Great Britain had been broken, did the Danish government representative veto the agreement, as the settlement of the Shetland and the Orkney boxes was clearly disadvantageous to the Danish fishing industry. At that time, the access problem had gained a prominence out of all proportion to its economic significance. Due to the existence of restrictions on the access to British waters, Denmark preferred *s* (the status quo) to *t* (TACs), and *t* to *u* (being left outside). Denmark ranked *s* first, then *t*, and then *u* (*s>t>u*), but in the end had to make a choice among the three alternatives. Although Denmark's order of preference was *s>t>u*, at the end of negotiations it chose *t*. What happened? The question here is why should an actor agree to a policy decision below his/her BATNA?

In this last bargaining round, the initial negotiating position of the Danish executive was to oppose any quota system imposing limits on the expansion of its fishing industry. Indeed one could argue that fisheries played a more important role in the Danish national income than in any other country of the EC[8], since Denmark is still the most important producer of fish within the EU in terms of volumes landed.[9] This, however, does not fully explain the Danish negotiating position. In order to understand why the Danish government made use of its veto power, the way the bargaining game was played between the EC and Denmark will be analyzed, and the domestic level will be taken into consideration for explaining the Danish negotiating position.

Figure 8 is a schematic representation of the bargaining process between the two players, EC and Denmark (DK). Denmark consists of two competing groups: DK_1, and DK_2. DK_1 stands for the Danish executive and DK_2 stands for the potential internal veto player the European Affairs Committee, representing the Danish opposition. To obtain a negotiating mandate DK_1 requires the support of at least a majority of the members of DK_2. In the negotiations with the EC, it is assumed that player DK represents the position of DK_1 and DK_2. On the other hand, the player EC represents the other member states and the Commission. Players alternate offers sequentially. One side (the EC) makes an offer that the other (DK) can accept or reject. The opening bid (the original position) gives a first impression of how far the actors are ready to compromise. If the offer is rejected, the second side (DK) can make a counteroffer of its own, and the first side can accept or reject that counteroffer. If the veto player accepts the offer made by the EC a new policy instrument can be adopted, called in the figure ANPI.

At the opening bid, the EC proposed a higher fish allocation for Denmark, but this was not sufficient to lift the Danish veto. Denmark wished to obtain 45 additional licenses for fishing in the area of the Orkney and Shetland boxes.[10] Denmark insisted on having access to these fish stocks and could even strengthen its negotiating position, because the nursery grounds of some important fish stocks (such as herring) lie predominantly within Danish waters. Thus, Denmark could threaten that in the case it were not guaranteed satisfactory access to the fisheries (for mature herring) in British waters, unrestricted fishing of the juvenile herrings would be permitted in Danish waters (Holden 1996, 237). The next offer made by the EC was to reduce the area covered by the Shetland and the Orkney boxes to three hundred square

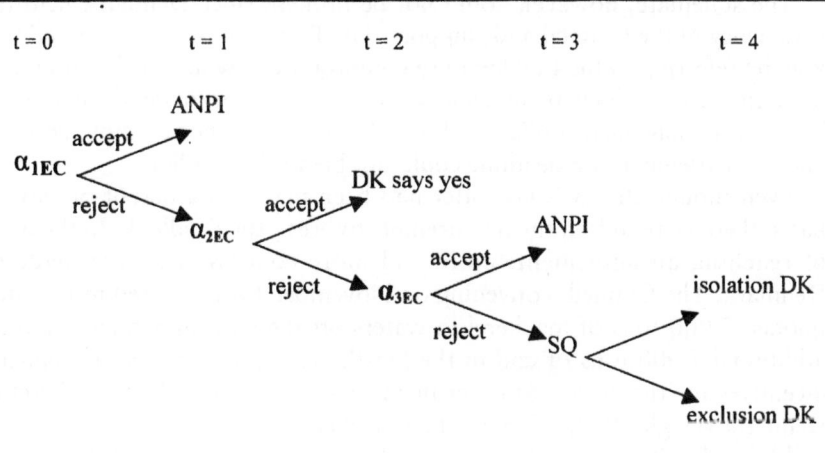

Figure 8: The extensive-form game between the European Community and Denmark

kilometers, a change in favor of Danish vessels. Denmark was still not satisfied with this concession, and increased its demand for seven licenses in the boxes and also requested additional quotas of mackerel from stocks west of Scotland. This request of the Danish representative was in turn unacceptable for Ireland and Great Britain (Leigh 1983, 87, 96).

The elapse of time led to the choice of certain strategies for trying to move the veto player to accept the side-payments. The Commission and the other member states threatened to leave the veto player outside the agreement and a deadline was agreed upon in order to break the stalemate. Introducing the possibility of threats during the negotiation process was the last resort and was used to elicit concessions. The central question here is whether the threat of being left outside was credible enough to move the veto player to agree. A threat is considered to be credible when players believe that sanctions (such as staying out of the agreement) will be inflicted. Thus the threat has to involve overt acts rather than intentions (Schelling 1960, 40).

The second strategy used was the settlement of a deadline with the expectance that, under time pressure, Denmark would accept the last compromise proposal that the EC had put on the negotiating table. Imposing a deadline is a way to exert pressure by stating openly the possibility of breaking up negotiations (Dupont and Faure 1991, 53). In this case, the deadline was set for November 5, 1982 by which date Denmark had to approve the last offer made by the EC. If the deadline were not met, the last compromise proposal of the EC would be withdrawn and the other nine member states would adopt and implement the Commission's proposals, bypassing Denmark's veto. To make the threat more credible, the British government, with prior consultation and approval of the Commission, prohibited Danish vessels from fishing within twelve miles around the coast of Great Britain.

The stalemate, however, could not be broken, since Denmark held the presidency of the Council and, supported by France, refused to proceed to a vote by referring to the Luxembourg compromise,[11] which at that time was often invoked to block an agreement. The threat of approving the conservation and management policy without Denmark was not credible enough, and the settlement of a deadline could not break the deadlock.

Even though the EC's last offer had been made on a take-it-or-leave-it basis, there were indeed further attempts to break the deadlock. In the hope of reaching an agreement, more and more concessions were made to Denmark. The Council, convening on November 1982, offered more catch quotas (7,000 tons of mackerel in waters off the west of Scotland and an additional 2,000 tons of cod in the North Sea) and a scheme of financial incentives for the vessels of other member states to land their mackerel in Denmark (Leigh 1983, 96–97). These new concessions to ameliorate what had been the EC's final offer, convinced opposition members of the Danish national parliament that further intransigence would elicit further concessions. Consequently, the Danish Social Democratic minority government refused the offer and a salami strategy was followed, that is to try to obtain one more slice (concession). Thus a new request was put on the bargaining table demanding that the proposed additional tons of fish have staying power, in other words continuity for more than one year.

Offers and rejections operate as signals to the other side and influence its willingness to accept or reject offers. In a negotiation process players may use concessions as a device to misrepresent their real preferences. Thereby the negotiator is kept in a dilemma: the more s/he concedes and thinks an agreement will be reached, as was the case for the EC, the further s/he will be from his/her initial negotiating position. A concession can be considered a sign of goodwill or as a sign of weakness. While in the first case it requires reciprocation, in the second case it elicits tougher behavior from the veto player, even leaving space for higher expectations (Dupont and Faure 1991, 46). What the EC considered as a sign of goodwill, in order to try to reach an agreement, was for the Danish veto player a sign of weakness.

At the Fisheries Council meeting held December 21, 1982 a further attempt was made to reach a compromise agreement. In order to facilitate the Danish parliament's approval of the final details, a direct telephone line was established between the Danish delegation at the Council of the EU in Brussels and the European Affairs Committee in Denmark. But not even this could move the Danish government representative to accept the compromise proposal. Once more the Commission in the name of the EC threatened to break off negotiations and exclude Denmark from the agreement. With the perspective that there might be yet another year ending without the fisheries conservation and management policy being settled, the Commission called upon the member states to implement its proposals by national measures. Moreover, it was under discussion that the Council might approve the CFP package by qualified majority.[12] At this point the negotiations seemed to have reached an impasse point, that is, the players had negotiated without reaching an agreement.

EXPLAINING THE POSITIONS OF VETO PLAYERS: DOMESTIC POLITICAL GAMES

When a member state uses the veto player card, negotiations in the EU must be modeled as an internal-external bargaining process, in which the national government representatives are constrained by their domestic level during the internal EU negotiation. The situations of veto players illustrate an executive's problems when playing external and internal games simultaneously. The negotiating position taken in international bargains is determined by the preferences of the executive and by the rules of the domestic political game. The reluctance of the three veto players on having a TACs system and a division of national quota at the European level is explained by domestic level factors, because a two-level game alters what Putnam (1988) called the domestic win set defined as the range of domestically acceptable solutions to an international agreement. Putnam (1988, 439–440) has proposed two hypotheses about the effect of the win-set on international negotiations: first, the smaller the win-set, the greater the risk that negotiations break down, and second, a small win-set can be a bargaining advantage. Adapted to EU negotiations, a similar hypothesis can be built: the bigger the win-set, the greater the possibility that involved parties reach an agreement.

In order to try to define which domestic level constraints were behind the negotiating position of the three veto players, it is necessary to go back to the way preference formation was defined in the conceptual framework for analyzing negotiations in the EU. It has to be checked which domestic factors can explain the negotiating position of member states for whom fisheries has an high salience. The focus shall be on time horizons, on the range of interest groups and their organizational effectiveness in the fisheries, the level of politicization of one issue, and on national parliaments as potential internal veto players.

The first factor constraining the negotiating position of the three veto players was their time horizon. Time horizon was a crucial variable for the British government in blocking agreement on the principle of equal access to fisheries resources. Before the legislative elections took place, the British government representative wishing to be re-elected, could not agree on an issue of disadvantage to 20 constituencies, which were also potential voters. After the British legislative elections in May 1979, the conservative government reformulated the British negotiating position without modifying the basic preference. The assumption is not made here that, with a new government, there was a shift in the British preferences. Aspinwall (2000: 415) points out, that since joining the EU, eight British governments under four leaders have always showed a similar degree of awkwardness towards the transfer of competencies to the supranational level. As the policy transfer on the conservation policy could not be blocked *ad infinitum*, at the beginning of a new legislative period the new British government had to define a price for its consent. It was prepared to change its initial position in demanding 60 percent of the total fish quotas, if some major restrictions on the access to British waters were settled. The result was the introduction of dominant preference for British fishermen around the Orkneys and Shetlands. No empirical evidence could be found for the importance of time horizon in the negotiating position of the Irish executive. The Irish veto blockade was essentially a bargaining strategy to obtain subsidies for its fishing industry. The Danish executive was constrained by its time horizon. The negotiations on the distribution of the national quotas occurred at the same time as the electoral campaign. It was not surprising that the Danish minority government was not prepared to agree to measures, which would have led to unemployment in the fisheries sector. It represents the typical case of an executive avoiding unpopular measures, because they would be linked to losing electoral votes.

The influence of national fisheries interest groups on their governments also explains why the veto player position was taken by the three countries. The absence of an agreement on the TACs had more advantages to all three veto players. Great Britain and Ireland preferred to keep the status quo, because they had rich fishing grounds. The pressure came essentially from British and Irish inshore fishermen. At the negotiating table, both countries demanded exclusive economic zones of 50 miles to protect the interests of their fishing industry. In the case of Denmark, the non-existence of legally enforceable quotas had led Danish fish industry to expand its fishing fleet, catches, and processing capacity and wished at that time to amortize its investments. Under such domestic circumstances any quota distribution, which would serve as a key for future years was not welcome. Therefore, the preference of Denmark was to have flexibility in the quota system with the prospect that the Danish quota could be increased in future years. In this way, the political benefits from the agreement were balanced against opportunity costs (the benefits foregone) and as the latter outweighed the former, there was without side-payments no incentive for the three countries to agree to something putting their fishing industry at a disadvantage.

Domestic salience and the level of politicization was high for the three fishing states, because the sector's economic and social role was important at the regional level. Job losses in this economic sector are politically highly sensitive, since the activities of the fisheries are concentrated within certain regions, often with no alternative employment. Furthermore, one job in fishing can lead to the creation of four to five jobs in processing, marketing, distribution, or shipbuilding. This explains why fisheries is considered to be a sensitive issue at the domestic level, why national government representatives bargained so hard when settling the TACs and national quotas, and how and why fisheries became a source of such disagreement in the EU. Contrary to the assumptions of neo-functionalist approaches that the European integration process would be gradual and at the beginning would focus merely on technical or non-controversial issues; sensitive or controversial issues were a part of the integration process from the very beginning.[13]

Finally, the role played by national parliaments was different for the three veto players in this distributive bargaining game. In Ireland, it played no role. In Britain, although there is a Select Committee on European Legislation in the House of Commons, which can submit any EU legislation that it deems important to the House of Lords for consideration, there was no empirical evidence that this committee tied the hands of the British executive. In the Danish case, an institution endogenous to the negotiations, namely the parliamentary oversight of the European Affairs Committee (EAC), constrained the Danish executive. Before going to negotiations at the European level, the Danish government always has to inform the EAC orally of its bargaining position. Thereby, the government can only start negotiations if this bargaining position is supported by a committee majority. This allows the EAC to bind the Danish government *ex ante*, and this is, compared to other European political systems, a unique situation. The weight of the EAC is especially high since Danish governments are often minority governments with automatically weak parliamentary basis.

During the bargaining process Denmark played the card of internal weakness in a credible way and could bargain with the EC from a position of strength.[14] This confirms Schelling's (1960, 28–29) conjecture about the power of weakness, in which he assumes that an executive can point to a recalcitrant legislature to extract greater concessions in international negotiations. Denmark claimed credibly that domestic pressures prohibited a disadvantageous compromise solution for its fishing industry. The Danish government representatives argued that the EAC, which provided the Danish minister with a negotiating mandate at the Council's negotiation table, would never accept the compromise formula. During the period in which Denmark adopted the veto player position in the conservation and management of fisheries resources at the supranational level there were from 1979–1981 and from 1981–1982 two Social Democratic minority governments (Green-Pedersen 2001, 10–11). The EAC constrained the executive's negotiating position on fishing quotas in 1982 and set the agenda for a compromise on the CFP in January 1983.

In other words, as the direct gains (the difference between the benefits expected from the policy transfer and direct costs) were negative, the three veto players refused to accept the proposal of the Commission on the conservation policy. Threats, offers, counteroffers and side-payments (financial aid for modernizing the fisheries fleet, restrictions on access to the fishery grounds and some supplementary fish quota) were necessary to lift their vetoes.

THE MEDIATOR ROLE OF THE COMMISSION: A BROKER IN ACTION

In a deadlock situation, the Commission can play an active and constructive role by acting as a broker and simultaneously seeking to find a final package deal that optimizes the joint gains of all the parties involved. During the distributive bargaining negotiations on the conservation and management policy, the Commission's role was more than that of a mediator between member states. It acted as a broker searching for compromise formulae to which all could agree. When the proposals of the Commission remained blocked at the Council during the last bargaining round with Denmark, it assumed an important role in persuading the other nine member states to adopt interim regulations. The Commission used the following strategy to implement this: it tried to persuade member states that they should agree to the proposed legislative framework with the prospect that this may be improved latter according to demands.[15] The Commission acted not only as a mediator between the parties, but also as a manipulator following its preferences when trying to move the parties towards an agreement.

The Commission while awaiting action by the Council of the EU, it ensured that member states adopted conservation measures in conformity with the EEC Treaty and with the objectives of the CFP. Furthermore, the European Court of Justice decided that the failure of the Council to act, did not imply that responsibility for conservation had reverted to the member states, who from 1 January 1979, had lost the authority to introduce new fisheries conservation measures in the waters under their jurisdiction. The Court, as an ally of the Commission in asserting and implementing Community responsibilities, was prepared to condemn member states if, by adopting measures which discriminated other member states, they failed in their duties under the Treaty. There were several Court cases on discriminatory national measures coming especially from Ireland and Great Britain. Thereby, the Court pronounced a verdict against both countries on the grounds of their discrimination of the fishermen of the other countries (Leigh 1983, 200–201).[16]

Driven by their own interests, supranational institutions are also actors in the negotiation process. This illustrates once more how agents seek to increase their power and are able to go beyond their formal competencies.

A SHORT ADDENDUM ON THE AUTONOMOUS PREFERENCES
OF SUPRANATIONAL INSTITUTIONS

If it is assumed that the preferences of supranational actors like the Commission are relevant, the question is how much autonomy do institutions really have in the negotiation process. One step further would be to ask how the balance between autonomy of the agent and its control by the principals can be evaluated. According to Pollack (1997, 108) delegation can entail two agency losses to the principals, namely agency shirking and agency slippage. Whereas agency shirking occurs when there is a conflict between the interests of the principals and those of the agents, agency slippage takes place when the structure of the delegation itself stimulates the agent to adopt a position different from those of the principals. A principal can always sanction an agent for not following the given instructions, but a central problem remains, namely that it is very difficult for a principal to prove that an agent has shifted the policy away from the principal's true preferences and towards those of the agent. Furthermore, even if there is evidence of agency slippage how can the Commission be prevented from going beyond its limits. Or as Buchanan (1975, 13) has put it: "how can the Leviathan be chained?"

McCubbins and Schwartz (1984) were the first to analyze the question of the autonomy of bureaucracies. They distinguished between two forms of oversight: police control and fire alarm, which subsequently were widely used in the principal-agent literature. More recently, Shepsle and Bonchek (1997, 369) went one step further and introduced a third form of oversight: the fire-extinguisher. While police control involves a centralized and direct surveillance, fire alarm merely requires a decentralized and indirect surveillance, for example interest groups and media report on the actions of agents. Finally, in the third form of oversight, the fire extinguisher, bureaucratic misbehavior is brought to the attention of the federal courts through service recipients. However, monitoring the agent's activities to determine the extent of agency losses is itself a difficult and costly task and also of limited effectiveness, since it depends on yet more agents or redundant bureaucracies (Bueno de Mesquita 2000, 101). Concerning the monitoring and sanctioning problems associated with the principal-agent relations, Pollack (1997, 107) points out that the degree of accountability of agents is difficult to measure. The reason why a principal is dependent on an agent is because s/he does not have the time and expert knowledge to deal with every issue. Consequently, national governments must rely on the specialized knowledge of agents, who are able to formulate a policy proposal. Thus, the agent can be considered to have some influence over the decision-making process.

The distributive bargaining game analyzed in this chapter has shown that in a stalemate situation, the Commission can make use of threats and deadlines in order to try to move the veto player to accept the side-payments. Furthermore, it acted basically as an intermediary between veto players and the other member states. The central issue was not that of controlling the autonomous preferences of agents, but rather to offer compensation (side-payments) to those member states that considered themselves disadvantaged by the initial proposal of the Commission on the fisheries conservation and management policy. Consequently, at this stage of the bargaining process on the settlement of the last pillar of the common fisheries policy member states were not concerned about agency losses, they just wished to find finally a compromise formula after seven years bargaining on the division of fish quotas among themselves. Further analyses focusing on the implementation process of the EC's legislation are necessary, in order to test how far the autonomy of bureaucracies goes and how member states behave when trying to monitor the activities of the agents.

EXPLAINING THE BARGAINING OUTCOME: SIDE-PAYMENTS AND THE SHADOW OF THE FUTURE

The negotiations on the conservation and management policy dealt with the distribution of fisheries resources among the fleets of member states. Veto players dominated the different bargaining rounds, which were influenced mainly by domestic political considerations in the three member states who had joined the EC in 1973 (Great Britain, Ireland and Denmark). Although the time pressure was considerable, negotiations advanced very slowly and decisions were postponed.

Initially, there were two veto players, Great Britain and Ireland, who maintained overall reserves on the unrestricted application of the equal access principle and demanded the inclusion of an extra criterion, namely contribution to the resources. Since the majority of EC's catches was taken in British and Irish waters, both players could strengthen their bargaining position. The veto players, however, had rich fishery resources in common, but heterogeneous preferences in fisheries. The central issue for Great Britain was to have an external policy securing fishing rights in Icelandic and Norwegian waters for its deep sea fishing fleet. On the other hand, Ireland was indifferent to this issue and was more interested in getting structural funds for the modernization of its fishing fleet. The consent of veto players was bought through side-payments: the settlement of fish boxes with restricted access (Irish, Shetland, and Orkney boxes), 36 percent of the total TACs for Great Britain, additional fish quotas and financial aid for Ireland.

After the initial reluctance of these two countries had been overcome, Denmark became a veto player in the last negotiating phase after the EC had agreed on restrictions to the principle of equal access. Negotiations were protracted, punctuated by threats, offers, counteroffers, and deadlines. Denmark bargained hard tending to resist making concessions and delaying agreement in the hope of achieving the best possible deal. This hard negotiating position also confirms Fearon's (1998, 282) affirmation, that in international negotiations the longer the shadow of the future, the more likely the use of a tough bargaining strategy is to be expected. As actors are bound for a long time by transferring a competence to the supranational level, before they agree to it, they will make sure that the settlement is not in conflict with their preferences. The negotiating position of the three veto players clearly demonstrates that in the EU, as in a cooperative game, each player calculates its benefits both with and without coordination. Thereby agreement can only be reached when all players expect their outcomes to be greater with coordination.

The decisive factor that obliged the Council finally to agree and to break the deadlock situation on the conservation policy was the perspective of the accession of the Iberian countries. The shadow of the future was looming once more over the actors. The Community needed a common basis an *"acquis communautaire"* especially with regard to the conservation and management policy before negotiating with Portugal and Spain, the latter was at the beginning of the 1980s in terms of landed volumes the third largest fisheries nation of the world. It was a mutual agreement that access to EC's waters and exploitation of resources should not be allowed to the Iberian fleets. Guaranteeing these two countries the full access to the provisions of the CFP would have wrecked it.[17]

After a long bargaining round the Danish consent could be bought with the allocation of some additional thousands tons of fish. In the future, Denmark would have special priority in the allocations of mackerel of up to 25,000 tons in Norwegian and Faroese waters (Leigh 1983, 98). But this concession implied that Denmark would give up its demands to a permanent share of the western mackerel stock. Finally, Denmark was also assured that its additional 2,000 tons allocation of North Sea herring would have staying power for at least three years. After the legislative elections had taken place and with these last minute concessions the new Danish right-wing minority government relented and accepted the side payments offered to obtain its consent.

The conservation and management policy was now ripe for a solution. As Zartman (1989) emphasizes a ripe moment is constituted by two components: on the one hand, the sides mutually experience a stalemate, that is they realize that nobody can benefit from a continued status quo; on the other hand, they see a negotiated solution as a possible way out of the deadlock (Jönsson 2001, 6). Game theoretic approaches assume that in a negotiation each player tries to resist making concessions, as s/he wishes to preserve his/her positions in order to achieve the best possible outcome. This is different in EU negotiations, where in the case of a veto player situation the EU, as an actor, tends to makes more concessions than a rational player would be expected to do.

After seven years of acrimonious negotiations, the framework for the adoption of TACs and quotas could finally be settled on the night of the January 25, 1983. All the players were exhausted by the long negotiations and were thankful it was over. Nevertheless, there was awareness that it represented an uneasy compromise.

The settlement of the conservation and management policy was compared to a "house of cards" (Holden 1996), which could easily collapse if any party tried to change one of the cards. It constituted a delicate balance of gains and sacrifices. Or as the President of the Commission at that time, M. Thorn, stated:

> Le cadre (de la CFP) a été mis en place au fil des ans pièce par pièce (...).
> D'y toucher serait terrible. Nous n'avons pas élargi le gâteau, nous l'avons

saupoudré et nous avons ficelé le paquet et la présentation est achevé c'est tout. Il y a toujours un moment où il faut mettre un terme à une négociation et je crois que ce moment est venu. Il y a des moments cruciaux si on rallonge la discussion on la fait capoter et on revient au point de départ et ce serait extrêmement dangereux.[18]

At the end of negotiations on distributive bargaining, member states decided that future decisions on fisheries issues would be taken on a qualified majority basis. The lesson from the seven years negotiations was learned: the unanimity voting procedure in fisheries issues was abandoned, in order to avoid future stalemate situations.

RELAXING THE UNANIMITY VOTING RULE: A SHORT INTERMEZZO ON SHOWING DISSATISFACTION WITH THE EU SYSTEM

When member states realize that under the unanimity rule the integration process stagnates, as any actor can manipulate the negotiating process, and the acceptance of a specific legislation is made dependent upon side-payments, they may give up the formal rule unanimity in exchange for QMV. The distributive bargaining game described in this chapter caused the ten member states to change the voting procedure accordingly. In this way, they wanted to avoid that in future the new accession countries, Portugal and Spain, would be able to block any reforms or proposals on the fisheries issue. In such a situation, actors can try to use their veto power in other policy areas to reach their targets. Again, the empirical research on fisheries issues in the EU provides a good example of how member states can push through their specific demands also after the unanimity voting on this policy field had been abandoned.

After the accession of Portugal and Spain, a distinction was made between the Community waters of the ten and issues concerning the two new member states, who were treated as outsiders. During the accession negotiations an extremely long transitional period of seventeen years was settled for the Portuguese and Spanish fishing fleets. From the very beginning both countries were most discontent with this compromise solution. Therefore, they were only waiting for the first opportunity to change their conditions of access in the fisheries sector. Thus, the question arises on how member states can show dissatisfaction with the EU system, if they do not perceive it to be in their interest, and if they cannot make use of their veto power. Theoretically two kinds of reactions are possible. First, member states may choose what Hirschman (1975) called the "exit" response. Discontentment is shown by disobeying the rules (for example, not implementing regulations). In most situations, however, rule busting is regarded as a violation of a norm and that may be punished by sanctions. Therefore, one would expect that member states also consider the second of Hirschman's options "voice", where they carry their disappointment to the forum.[19] This was what Portugal and Spain did, when they showed the forum their disagreement through their uncooperative behavior, with reservations on all fisheries issues discussed.[20] Drawing on this Grieco (1996, 288–289) asserts that when new international institutions are being negotiated, states will make sure that any cooperative agreement will include effective voice opportunities. Grieco hypothesizes that if such voice opportunities are absent, actors will try to renegotiate the terms of the institutional arrangement, and furthermore they may reduce or withdraw their commitment to the organization in the case when such attempts prove to be unsuccessful.

After being unable to block the reform of the CFP in 1992, since decisions were reached with QMV, Spain threatened to veto the accession of Sweden, Finland, and Austria to the EU, if the transitional measures for Spain in the fisheries policy were not abolished. In order to avoid a deadlock situation, the date for full accession to Community waters was advanced to 1996, and 40 Spanish vessels were allowed to fish in the Irish Box area (Lequesne 2001: 37).[21] This case illustrates how a new member state can take the first suitable opportunity to renegotiate the conditions of access in the fisheries policy and also demonstrates that abandoning unanimity voting in one policy issue, does not eliminate a veto power of a certain member state on this subject, since the issue can be linked to other issues subjected to unanimity rule. It shows that when new member states obtain power, that is are a part of the "club", they will try to change the rules through renegotiations and veto threats over issues. In this way, restrictive measures have to be abolished, otherwise the EU will find itself deadlocked.[22]

CHAPTER 5

Conclusion
Drawing the Threads Together

THE THREE FINDINGS

The primary aim of this dissertation project was to comprehend how negotiations in the EU work theoretically and empirically and to do this a conceptual framework for analyzing internal negotiations was developed and applied to two key negotiations in the settlement of the CFP, which at the same time illustrate how integrative and distributive bargaining takes place in the EU. It was argued that preferences of member states are an important variable when assessing EU negotiations. However, taking preferences seriously is not the whole story of bargaining in the EU. In order to be able to explain bargaining outcomes, not only the preferences of member states, but also the preferences of the European Commission and the institutional setting have to be taken into account. This work attempted to analyze negotiations in the course of the European integration process by linking theories to reality. An analytical framework based on the tradition of rational-choice was used and state-centric approaches were linked to new institutionalist and public choice approaches. The focus was placed on explaining EU negotiations and bargaining strategies specifying who were the actors at the negotiating table, which preferences they had, and how far the institutional setting and the preferences of the Commission influenced bargaining outcomes. It was shown that for explaining bargaining outcomes in the EU, not only the preferences of the member states, but also the preferences of the Commission and the institutional setting played a decisive role. In the two analyzed bargaining situations it was demonstrated how these variables affected the bargaining outcomes. Three findings come out: *preferences of member states matter, the preferences of the European Commission matter*, and *the institutional setting matters*.

PREFERENCES OF MEMBER STATES MATTER

The first finding of this book refers to the general statement that the preferences of member states matter. Integrative bargaining in the EU has shown that:

1. When there are strongly polarized preferences — some member states are in favor of issue x (here structural policy), but against issue y (common market organization), while the preferences of the other players are in the opposite — coalition-building and issue linkages are the negotiating tools under unanimity voting for reaching agreement.
2. Actors with net gains from a European policy will favor transfer to the supranational level, whereas actors being net contributors to the EU-budget are rather reluctant supporters of any new initiative inter-linked with new financial costs.

In the integrative bargaining game the intensity of the preference for the structural policy was very high for two member states, but low for the remaining four member states. In contrast, in the case of the common market organization four members states had a very high preference, whereas the preferences for the two other actors were low. When the preferences of member states are strongly polarized, the number of actors involved is small, and two policy-issues are bargained, the building of coalition and the linkage of issues are the negotiating tool used in negotiations for achieving mutually beneficial outcomes.

The formation of two coalitions allowed issues to be linked. *Coalition A* constituted by France and Italy wished to have a structural policy in fisheries totally financed by the EC. Since the level of openness of the markets for fish products would increase, *Coalition A* would have to modernize its production structures and demanded a structural policy to equilibrate the costs linked to a change of the status quo. *Coalition B*, formed by Germany, the Netherlands, Belgium, and Luxembourg wished to have a common market organization for fish products and was not willing to pay the largest share of the new system benefiting *Coalition A*. While *Coalition A* was primarily concerned with getting subsidies to modernize its fisheries fleet, *Coalition B* was trying to avoid the settlement of a further interventional policy and automatically a further increase in its contribution to the EC's budget. Thus the bargaining outcome constituted a compromise between the positions of both coalitions: a partial liberalization of fish markets inter-linked to withdrawal prices and a structural policy financed to fifty percent from the EC-budget. *Coalition B* accepted the demands of subsidizing the modernization cost of *Coalition*'s A fishing fleet, because in return it would obtain a trade liberalization of fish products. It was a typical situation of bargaining in the EU: member states exchanged concessions in one issue-area where their preferences were relatively weak for concessions in another area where the salience of an issue was higher. This package deal created gains for all involved actors and an agreement that was Pareto-superior to the status quo could be reached. It represented a win-win outcome, a positive sum issue linkage with mutual advantages for all parties involved. Since member states did not have to make use of their veto power, negotiations took a relatively short time.

In the distributive bargaining game dealing with the distribution of fisheries resources among members and how the preferences of the member states mattered, following findings can be made:

1. When there is an unequal preference distribution (for some actors the domestic salience of the issue bargained is low, while for others it is high), and only one-policy dimension is bargained, those actors who have a high salience on the issue bargained, dominate the negotiations through their veto power and can push through their preferences.
2. The weaker a national government is internally, the more reluctant it will be in accepting those European policy measures that would have negative consequences at the national level.
3. The weaker a national government is due to domestic pressures, the stronger is its relative bargaining position in EU negotiations.

Negotiations on the conservation and management policy were characterized by two distinct phases: in the first bargaining round, there were two veto players, Ireland and Great Britain, and the negotiations dealt with the questions of the unrestricted or restricted application of the principle of equal access to resources and on how the total allowable catches should be divided among member states. In the second and last negotiating phase, the Danish government vetoed the agreement on the conservation and management policy. These two apparently technical rubrics (the application of the principle of equal access and TACs) camouflaged a political problem of resource distribution. The negotiations about the distribution of fisheries resources among member states were like stirring up a hornet's nest. The three veto players saw their interests better served by maintaining the status quo, and accordingly during the negotiations they vetoed the Commission's proposal on the conservation and management policy. After seven long years of negotiations, the consent of veto players could be bought through side-payments: settlement of fish boxes with restricted access (Irish, Shetland, and Orkney boxes), 36 percent of the total TACs for Great Britain, additional fish quotas and financial aid for Ireland and Denmark.

The analysis of this second negotiation on distributive bargaining demonstrates that in the case of uncertainty about the chances of a final agreement, actors, who really desire an agreement, will be willing to sacrifice relatively more than those actors who are not so dependent upon an agreement. In game theory terminology: the player, who is less willing to risk a breakdown, will make a concession to the other (Zeuthen principle). Actors with greater bargaining power, such as Great Britain, Ireland, and Denmark, determined

the policy outcomes, whereas weaker actors, in this case the remainder countries of the EC, had to make side-payments to the veto players in order to
reach an agreement.

In order to explain why member states make use of their veto power, the
factors used in the definition of preferences of member states in the conceptual framework were juxtaposed to the reality. *Table 5* shows how strong or
how weak the major determinants in the preference formation of member
states were for Ireland, Great Britain, and Denmark.

Time horizons of political decision-makers were weak for the Irish, strong
for the British representative and very strong for the Danish representative.
Before elections take place, national government representatives strongly
defend their preferences and therefore tend to block agreements. This was the
case for the British government and even more so for the Danish minority
government. It was easier to get a government to agree to side-payments at
the beginning of a legislative period than before elections took place. This
was corroborated by the British and the Danish bargaining strategies. After
the elections took place, both governments accepted the side-payments
offered to them by the EC and an agreement could be reached. Moreover, the
organizational effectiveness of interest groups at the national level and also
the degree to which an issue is politicized in a country was very high for the
three veto players. At the national level, inshore fishermen who expected to
lose from a change of the status quo mobilized and put pressure on the
national governments. The absence of an agreement on the access to fisheries
resources and how they should be divided had more advantages for all the
three veto players. In the case of Great Britain and Ireland it was because they
had rich fishing grounds, and in the case of Denmark, because the fish
processing industry had just modernized its fleet and wished to expand its
catches. The level of politicization and the domestic salience of fisheries was

Table 5: Distributive bargaining game with veto players

Preferences	Veto players		
	Ireland	Great Britain	Denmark
Time horizon	–	+	++
Organizational effectiveness of interest groups	++	++	++
Level of politicisation of one issue	++	++	++
National Parliaments as veto players	–	–	++

Legend of the figure: — *weak*
              + *strong*
             ++ *very strong*

very strong for Great Britain and Denmark and also strong for Ireland, since accepting total allowable catches and an unrestricted application of the equal access principle would lead to losses of employment to their inshore fish-ermen. Due to the concentration of fisheries activities within peripheral regions, job losses in the fisheries sector were considered politically highly sensitive. Finally, national parliaments as internal veto players constituted the last indicator for explaining why member states make use of their veto power. The power of the national parliaments to constrain the executive at the supranational level varies from country to country and depends on the constitution and political system of each member state. In Ireland and in Great Britain there was no empirical evidence for the role of national parlia-ment in constraining their negotiating positions. Only in Denmark the national parliament with its European Affairs Committee, which had to give a negotiating mandate to the Danish minority government, acted as a hand-tying institution endogenous to EU negotiations. This allowed the Danish government representative to use credibly the "paradox of weakness" as a negotiating strategy. It could play the role of a weak negotiating player constrained strategically by its own bargaining space for a viable agreement. Denmark claimed that domestic pressures made impossible the acceptance of a disadvantageous compromise solution for its fishing industry. When an actor is constrained domestically or can persuade its counterparts that s/he is constrained at the domestic level, the EU has to make more concessions to that actor. Hence the greater domestic constraints are, the more they can enhance a veto player's bargaining power. In the last bargaining round of the negotiations on the settlement of the conservation and management policy, the Danish veto power clearly shows that the actor who is willing to wait longer, and who is less eager for an agreement to be reached, will have more success in maximizing his/her preferences.

PREFERENCES OF THE COMMISSION MATTER

The second finding that preferences of the Commission also matter in nego-tiations in the EU could be demonstrated by analyzing the two different bargaining games. The formal agenda-setting power of the Commission was an important factor for pushing through its preferences enabling the Commission to influence policy outputs. The preference of the Commission is not only to expand the scope of Community competence in a policy field, but also to increase its own influence within it. The decisive factor for an actor like the Commission is the settlement of policy instruments that can be monitored at the supranational level.

Under unanimity voting the Commission is able to exploit the differences between member states' preferences and thus to introduce its own prefer-ences into a policy proposal. When agents do not share the same interests as the principals, the agent is in a strong position to tilt choices away from the principal's preferences, trying to shape policy proposals to fit more closely its own objectives rather than those of the principals. Empirical evidence for this affirmation is given in this dissertation. In the integrative bargaining game at the beginning of the negotiation process, the Commission could expand its formal agenda-setting power. When the Commission was asked by the member states (France and Italy) to prepare a regulation on the structural policy and the common market organization (taking into account the demands of Germany and the Netherlands), the Commission used this window of opportunity and included elements for a general common policy in the proposal. It constituted a unique opportunity to broaden its competencies through a policy transfer from the national to the supranational level. The Commission manipulated the agenda setting process and acted as a policy entrepreneur through innovative policy proposals. This clearly illustrates the situation in which member states instruct the Commission to prepare a text, but its content can no longer be controlled by the principals. While the member states, with their short time horizon, wished to have some measures on a certain policy issue, the Commission, with its long-time horizon, was already manipulating the threads for having a competence transferred to the European level.

In the integrative bargaining game the focus was on coalitions between member states, whereas the distributive bargaining games was centered on veto players. This allowed one to test the role played by the Commission under these two different bargaining situations. Under a coalition-building situation, the influence of the Commission during the bargaining process was rather weak: it had a mediator function in linking issues and tried to find a compromise formula which could be accepted by all involved parties. The Commission acted as a mediator between member states, since it could devise new compromises. It is also in the Commission's own interest to keep negotiations moving and to exert some pressure in order to obtain a settled agreement.

In the distributive bargaining game, the primary aim of the Commission was to have a centrally monitored system for dividing the fish stocks among member states. As a result, the proposals of the Commission on the conservation policy concentrated upon the sharing out of available fish stocks through total allowable catches and a national quota system. The choice for this policy instrument was motivated by utility maximizing considerations, since this management system had to be centrally monitored and enforced allowing the Commission to expand its power and deepen the European integration process. This illustrates once more how the formal control of the agenda setting gives the Commission the opportunity to act strategically and to achieve a system more in accordance with its preferences. Moreover, when there was a veto player situation in the distributive bargaining game, the Commission could have a stronger influence on the negotiation process than when coalitions were built. The Commission not only kept the negotiations going on by drafting and redrafting proposals, but also assumed an important role by isolating veto players and was able to take over the role of a broker. In order to try to bypass the Danish veto, it proposed to take a vote in the Council under qualified majority voting. In this way, Denmark would have been isolated. The exclusion threat was used in order to try to move the veto player to accept the side-payments and to lift his/her veto. The Commission also did not act as a simple agent waiting for the principals to give it a task. Since the Commission's compromise proposals remained blocked in the Council, it assumed an important role in persuading the other member states to implement other aspects of a common policy, which would subsequently be approved. Driven in the pursuit of its own preferences, the Commission, as an actor at the negotiating table, had an important role in trying to move member states toward an agreement.

Finally, it was found that the European Commission can move the system and influence decisions, when it has the support of some states. Joint decisions have been produced far beyond the lowest common denominator (the expression of the position of the most reluctant actor), of what member states have originally wished. The analysis of the two negotiating situations demonstrates that under unanimity voting the Commission can act like an autonomous actor. An astute mediator, like the Commission, examines those factors that A and B can accept and generally proposes a compromise solution, in which the best of both worlds are combined. The Commission acted rationally as a purposeful opportunist using every window of opportunity to expand its policy domains and to create new ones.

THE INSTITUTIONAL SETTING MATTERS

The third contention of this analysis is that the institutional setting also matters in negotiations in the EU. The empirical analysis has shown how it played a role and which bargaining tools were used to overcome the unanimity trap.

In the integrative bargaining game, it was not difficult to overcome the unanimity rule, as coalitions were built, issues could be linked (common market organization with structural policy) and a package deal tied, with something for everyone. A strong preference relation existed, all actors placed great value on the issue discussed: some member states had a strong preference relation concerning the liberalization of markets for fish products, while others put a high value on having a structural policy financed entirely by the Community. Due to the existence of these opposed and polarized preference relation, the costs of linking issues were not too high and these two policy dimensions could be linked together facilitating agreement. Iterated bargaining as an intervening variable played a role: the shadow of the future (the foreseen first enlargement of the EC) gave the impetus to come to an agreement before the new member states, which were also key fishing states, adhered. Delaying agreement meant automatically that preferences of more actors would have to be taken into account in the next bargaining round or eventually that the negotiation would have to start from zero.

In the distributive bargaining situation, one side (at least one actor), using the formal rule unanimity voting, could manipulate the bargaining process to his/her advantage. The strategy of the veto players was to veto an issue temporarily and in this way increase their power. It can clearly be corroborated that the unanimity voting rule places each actor in a position where s/he can bargain bilaterally with the other member states and with the Commission from a position of strength. This feature explains why the costs of reaching agreement under unanimity can be extremely high. The informal rule, iterated bargaining, played a role during the negotiation process as an intervening variable, since only the shadow of the future, the accession of Portugal and Spain both with strong interests in fisheries, gave the final impetus to reaching an agreement. After having been negotiating seven long years on the CFP, the decisive factor that obliged the Council to act was the perspective of the accession of the Iberian countries. Furthermore, the EC had suffered from the negative experiences with new members (Great Britain, Ireland, and Denmark) that persistently blocked any introduction of policy measures to introduce the conservation policy and introduced qualified majority voting in fisheries. It was the simple concern about the future that fostered cooperation. When repetitions of a negotiating game are foreseen for the future, every move is calculated with reference to the opportunity costs associated with the next interaction.

Table 6 compares the two analyzed bargaining games under unanimity voting, which was used as a constant parameter to test whether it may have different effects on bargaining outcomes. It turned out that unanimity voting has indeed different effects. In the integrative bargaining situation the number of players involved were small (six), preferences were highly polarized, two-policy issues (structural policy and common organization of the market) were under discussion and issue linkage was used as a negotiating tool.

Adding or subtracting players into a game's configuration allows one to analyze how easy or difficult it is to reach an agreement under unanimous decision-making. It is expected that the larger the number of players, the more difficult it should be to reach an agreement and to link issues. In practice, however, this general rule may not necessarily apply. This may be the case when member states fall into blocks with similar preferences, with the result that the bargaining process becomes easier and shorter. The conceptual framework predicted that adding more actors (member states) should create pressure for a change in voting rules. Furthermore, what matters is not how many countries are added with the next enlargement, but rather if they have weak or strong preferences on a certain issue. In the distributive bargaining situation, the number of players was increased (from six to nine) and the intensity of preference was low for the majority of member states and high for the minority, who placed great value on the issue discussed and made use of their veto power. Since only one issue (here, conservation and management

Table 6: Comparison of two bargaining games under unanimity voting

Empirical Field of Research	Preference Intensity	Number of Players	Policy Dimension	Negotiating Tools	Outcomes
Integrative bargaining	Strong	Six	Two-policy	Coalition-building + Issue linkage	Package deals
Distributive bargaining	high and low salience	Nine	One-policy	Veto players	Side-payments

policy) was bargained, veto players dominated the different bargaining rounds. When all these conditions are fulfilled, only one-policy dimension is under discussion, a high domestic salience of an issue for a minority and a low domestic salience for a majority and unanimity is the voting rule, agreement can only be reached by buying the consent of the veto players by side-payments.

SOME GENERALIZATIONS ABOUT THE NEGOTIATION PROCESS IN THE EU

A number of generalizations about the negotiation process in the EU can be made in concluding this study:

1. The analysis has demonstrated that negotiations in the EU under the fist pillar of EU-governance and falling under the consultation procedure, constitute the sum of the preferences of the actors involved (member states and European Commission) plus the institutional setting, constituted by the formal (voting rules) and informal rules (iterated bargaining) of the game. The transfer of a competency to the supranational level can be achieved only when the expected utility of an agreement is bigger than the eventual losses for a single actor.

2. Last-minute compromise is a further feature of the EU negotiations. A compromise is reached in extremis at the end of intense iterated negotiations. Compromise means here that in the multilateral negotiations actors try to find an agreement, which involves mutual concessions (issue linkages and side-payments). When some member states and the Commission prefer a decision to a deadlock, the next step is to try to find incentives for bargaining and the exchange of (mutual) concessions. As the empirical analysis demonstrated, the negotiation deadlock in the EU can be broken, when it is possible to offer compensatory rewards to the veto players. Thereby, agreement depends on a cost-benefit calculation:

if the benefits surpass the opportunity costs, that is remaining outside the "club's" new policy, the reluctant member state(s) will accept the offer made by the other member states and the Commission.

3. The bargaining outcomes that are finally reached represent the result of a series of bargains, concessions, package deals, and internal linkages. The EU negotiation process is characterized by delicately balanced compromises leading to a strong rigidity in the system. It can be considered as a house of cards, if you try to change one element (a card) all the others may fall down.

4. In dealing with the conservation policy and management issues, especially the settlement of national quotas, the concern of each member state was clearly to maximize its own benefits and minimize its own losses. Many of each national government representatives went into maximizing the short-term opportunities for his/her national fleet with little apparent regard for longer term consequences or wider conservation issues. There was very little sense of sharing responsibility for a common resource. The protection of fish stocks was never a priority in the negotiations.

5. Under unanimity voting the larger the number of actors involved in the bargaining process, the more difficult the decision-making process is. In the first CFP package the homogeneity of the actors was greater, thus reaching agreement was easier and the costs of arriving at decisions lower through coalition-building and issue linkages. With the first enlargement, the number of the actors involved in the bargaining process doubled, consequently the heterogeneity of the club was increased and so the costs of reaching agreement.

Neither the liberal intergovernmentalist assumption that national preferences determine EU bargaining outcomes nor new institutionalist approaches, which assume that the institutional setting determine bargaining outcomes are correct. The difficulty is that the truth falls somewhere between these two approaches: bargaining outcomes in the EU are the result of the preferences of the member states, and the preferences of the Commission added to the institutional setting. The focus of this thesis was on the Council-Commission's tandem and only bargaining outcomes involving these two institutional actors were considered. The conceptual framework does not pretend that the preferences of the Commission and the institutional setting determine EU negotiations, it only tries to find out what difference they make. The negotiations on the establishment of the CFP clearly demonstrate that negotiation outcomes in the EU are a hodgepodge of compromises, the result of a power struggle between preferences of member states, preferences of the Commission and the institutional setting.

The integrative and distributive bargaining situations on the settlement of the CFP illustrated a situation in which actors preferred coordination to noncooperation, but where they had different policy positions over the outcomes. As the EU is a forum for bargaining, all the players try to reach an agreement where they can see themselves as "winners". Most of the time in the EU, we have what Schelling (1960) called a prominent solution to a coordination problem, in which the important thing for actors is that they agree on something, and what is agreed on is secondary. In other words, in the EU reaching agreement on a common policy has greater priority than finding the "best policy". The bargaining process illustrates clearly how the EU operates. Bargaining outcomes give something to everyone and the preferences of all the participants are taken into account in the policy formulation creating complex, bureaucratic texts which in the end constitute mere hollow words.

The threat of veto remains a useful bargaining tool in negotiations in the EU through which veto players try to obtain as many concessions or side-payments as possible. Unanimity voting creates crippling opportunities for strategic manipulation. During the negotiations on the distributive bargaining, veto players made use of their veto power to block a decision. When an actor wishes to maintain rather than to change the status quo, under the unanimity rule any actor can manipulate the negotiations until an agreement is reached which is in accordance with his/her preferences. After seven years bargaining on settling the access to fish stocks and the division of the total allowable catches, member states decided to exchange the unanimity voting procedure for qualified majority voting in fisheries in order to avoid future stalemate situations. In the EU, where not only internal but also external issue linkage is possible, the problem, however, remains. When member states cannot make use of their veto power in a certain policy issue they will make use of it in other areas under unanimity in order to reach their targets. Abandoning unanimity voting in one policy issue does not eliminate a veto power of a certain member state on this subject, since the issue can be linked to other issues subjected to unanimity rule. Such situations can only be avoided by relaxing universally the unanimity voting procedure.

Notes

INTRODUCTION

1. Whenever referring to the EU after the Treaty on the European Union came into force, the term EU will be used and for the period before, the expression European Community (EC).
2. When referring to the European Commission, the expressions European Commission and Commission will be used interchangeably.
3. When analyzing the European integration process, the first problem one encounters is that there is no clear definition of what is meant by European integration process. Haas (1961, 367–368) was one of the first to provide a definition. He understood integration to be a " *process whereby political actors in several distinct national settings are persuaded to shift their loyalties, expectations, and political activities toward a new and larger center, whose institutions possess or demand jurisdiction over the pre-existing national states."* Moravcsik (1998, 5) defines integration as a process in which the governments *"define a series of underlying objectives or preferences, bargain to substantive agreements concerning cooperation, and finally select appropriate international institutions in which to embed them."* For Pfetsch (1998, 303) and Hix and Goetz (2000: 3) the European integration process is a synonym of the *"pooling of sovereignty"* which comprises two aspects: the transfer of competencies to common institutions with executive, legislative and judicial powers (supranationalism) and the process of coordinating different national interests.
4. Wallace (1998, 42) distinguishes further between the following policy modes: partnership, segmentation, co-option and cooperation. Each one of these policy modes is particularly associated with different periods in the evolution of European integration process and all of them are still present in the discussion of the further development of the EU. Peters (1992) defines EU governance as "bureaucratic politics" and Majone (1994) understands it essentially as "regulatory politics."
5. Note that *"polity"* is defined as a system of governance capable of producing decisions.
6. There is a broad range of negotiation definitions. Rubin and Brown (1975, 2) were among the first to state that bargaining or negotiation is a *"process whereby two or more parties attempt to settle what each shall give and take, perform and receive, in a transaction between them."* A further definition was given by Lax and Sebenius (1986, 11), who understood negotiation as *"an*

attempt by two or more parties to find a form of joint action that seems better to each than the alternatives." Moravcsik (1993, 497) defined negotiation as *"the process of collective choice through which conflicting interests are reconciled."*

7.   In January 1958 the Ministers of Foreign Affairs created a Committee of Permanent Representatives (COREPER) which was given the task of preparing debates and managing daily routine (Westlake 1998, 286). For a detailed analysis on this quasi-institution and how it maintains the performance and output of the Council through the production of a culture of compromise see Lewis (1998).

8.   The term *integrative and distributive bargaining* was introduced by Walton and McKersie (1965) for analyzing labor negotiations.

9.   In European studies, the term *bounded rationality* was introduced by Bueno de Mesquita (1997, 236) referring to individuals who *"are not able to look ahead over an unbounded time horizon."*

10. Bazaar tactic or technique is a metaphor for negotiating tactics demanding an exorbitant price for the consent of a certain player.

11. This study intends to concentrate on the use of the game theory in bargaining, which can also be applied to other issues in political science, such as the role of legislative rules, voting in elections, etc. (cf. Morrow 1994, 2–3).

12   For a recent and general discussion of the various new-institutionalist schools and an assessment of their relevance and limits to the current state of European integration see Hall and Taylor 1996; Kato 1996; Jupille and Caporaso 1999; Aspinwall and Schneider 2000; and Peters 2000.

CHAPTER 1: THE SETTLEMENT OF THE COMMON FISHERIES POLICY

1.   Presently, the Council fixes the TACs for more than 120 fish stocks from the 224 existing fish species in EU waters. Thus there is a difference between the so-called "analytical TACs" based on solid and sufficient scientific data and the "precautionary TACs," calculated by using more or less ambiguous methodologies. The evaluation of fish stocks covered by TACs is based on an arrangement, which classifies them according to one indicator: the effect of fishing mortality on the average yield condition. Although this criterion is somewhat arbitrary and uncertain, it shows that about 77 percent of the analytical stocks are either fully exploited or depleted (Karagiannakos 1996, 236–238; Lequesne 2001, 88).

2.   Technical conservation or gear restriction measures were settled in order to influence the catches pulled out of the sea. Its method consists in imposing restrictions on the types and specification of equipment used. A distinction was made between the *minimum mesh sizes* and the *minimum landing sizes*. The first is very difficult to operate in multi-species fishery and the second impossible to enforce, unless inspectors are prepared to control each fish market (cf. Coffey 1995, 16; Gray 1998, 7).

3.   The "Hague preferences," laid down in 1976, refer essentially to the necessity of taking into account the vital need of local communities and industries dependent on fisheries when implementing the CFP. Thus a socio-economic element was incorporated into the process of allocating fishing rights.

4.   During the 1994 accession negotiations, the state of fisheries was one of the major issues for Norway, essentially because the fishing industry contributes to six percent of the country's exports. From a Norwegian perspective accepting the CFP had some disadvantages. First, Norway, has a good conservation system and membership to the EU threatened to weaken this national system of

fisheries management. Second, the fisheries issue was used to benefit the existing members of the EU, it would have offered tariff free access to Norway's most important market (Sœter 1996,144–145).

5. The UNCLOS divided maritime waters into different functional areas: *internal waters* give each coastal state the right to establish a territorial sea up to 12 miles from its baselines; *archipelagic waters* extend within the baselines joining all the outermost points of the islands and dying reefs of the archipelago; *continental shelf* comprises the sea-bed and subsoil of submarine areas extending beyond the territorial sea of coastal states throughout the natural prolongation of its land territory to the outer edge of the continental margin or to a distance of 200 nautical miles from the baselines (Garcia and Hayashi 2000, 452). The high seas remain free. For a more detailed history and evolution of the international law of the sea, see Garcia and Hayashi (2000), Hanna (1998), Rettig (1996).

6. "I think we should not forget one important point. In the very difficult situation of extending the fishing zones to 200 sea miles, the settlement of a common fisheries policy puts us in a position where we can exert great economic and political influence on international fishing jurisdiction. As a single member state we could not do this and would be at the mercy of the large coastal states, this is what should inspire us." (translation by the author). Statement of the German Representative. Quotation from the stenographic record of the *Minutes of the Fisheries Council Meeting* from 25.01.1983, Bobbin No. 2596, p. 12.

7. There is now a broad discussion on absolute and relative gains in international relations theory. While neo liberal institutionalism supposes that states focus primarily on their absolute gains and the prospects for cooperation are emphasised, structural realism assumes that states are concerned with relative gains and emphasise the prospects for conflict. For a detailed overview see Powell (1991), Snidal (1991), and Berejekian (1997).

8. For a general framework on how to analyse public goods see Oakerson (1992) and Weimer (1997).

9. There is a broad literature in this area, for a general overview see Bromley and Cochrane (1996), and Oakerson (1992).

10. Biological studies, however, like that of Ávila de Melo and Alpoim (1999, 7–9) emphasise that the collapse of the Grand Bank stocks is the result of the accumulated action of three different factors: overfishing due to the lack of an effective gear and effort control; extreme climate and oceanographic conditions during the late eighties and early nineties; and the continuous growth of the harp and hooded seals stocks and their concentration southwards, overlapping with the Grand Banks area. Furthermore, this study emphasises that it is also important to differentiate which cod stocks are analysed. The same authors come to a different conclusion in relation with the collapse of the Flemish Cap cod. In this case, the collapse is a direct consequence of the systematic overfishing of this stock during the 1985–1995 period. For an overview of fisheries management in the Grand Banks see Mitchell (1997).

11. The quota hopping system consisting in a system that permits nationals of one member state to buy vessels of another member state and to register them there in order to obtain a share of the quota allocated to that country is another contentious issues in the CFP, as it encourages fishing up to the limit of the quotas and at the same time leads to high rates of fish discards and it represents a temptation for fishermen to sell over-quota fish on the black market.

12. Decommissioning refers to the financial compensation paid to fishermen for withdrawing their vessels from fishing. However, because the compensations are generally low its effect on fishing is rather small (cf. Gray 1998, 7).

13. Some studies focus on the states' performance in implementation (Garza-Gil et al. 1996; Steins and Edwards 1997; Thom 1999), while others analyse the issue of compliant vs. non-compliant behaviour of fishermen (Honneland 1999). As Gray (1998, 8) argues the management failures to implement fishing regulations is also the object of controversy, mainly because the costs of effectively controlling the fishing industry would be enormous. The task of monitoring adherence to mesh sizes, landing sizes and quota regulations (including by-catches) is practically infeasible in the EU with its hundreds of fish markets and tens of thousands of fishing vessels. If policing measures were really implemented, there would be a disproportionately high expenditure on equipment (including high-powered patrol boats), surveillance and labour.

14. It goes beyond the purpose of this work to distinguish between the different bio-economic models and their evolution in the last fifty years. For a summary discussion, see Anderson (1996), and Flaaten et al. (1998).

CHAPTER 2: CONSTRUCTING A SPRINGBOARD

1. The fist pillar refers, in EU jargon, to the European Economic Community (EEC), to the European Atomic Energy Community (Euratom) and to the European Coal and Steel Community (ECSC). All the Communitarian policies that are associated with the Single Market fall under the first pillar, that is policies in which a policy competence has been transferred to the supranational level, like agriculture, fisheries, competition, environment, energy, economic, regional, commercial, structural, and economic and monetary policy.

2. In the EU there are essentially four legislative decision making procedures: the consultation procedure, the cooperation procedure, the assent procedure, and the co-decision procedure. For a detailed analysis of these different procedures see Steunenberg and Selck (2001).

3. Under the consultation procedure the Commission submits a proposal to the Council and asks the EP for its opinion. In a next step, the EP gives its opinion, the Commission may amend its proposal on the basis of the EP's opinion, but it is not obliged to do so. Finally, on the basis of unanimity or qualified majority voting, the Council may amend or adopt the proposal.

4. It must also be pointed out, that in the meanwhile, a number of papers have appeared focusing on the EU Council Presidency, analysing its agenda-shaping powers (Tallberg 2001) and as a mediator (Elgström 2001; Beyers and Dierickx 1998). For the sake of simplicity the analysis shall be limited to the Council-Commission tandem.

5. For a general overview on political systems in western European countries and especially on the extent of centralisation or federalisation of legislative authority see Conceição-Heldt (1998).

6. Three key fishing countries may illustrate this diversity. In Spain, there are *Cofradías* (fishermen's guilds), which are grouped into regional federations, Shipowner's Associations and Business Association and Trade Unions. In Great Britain, there is on the one hand, the *National Federation of Fishermen's Organisations*, which represents interests in England, Wales, and Northern Ireland, and on the other hand the *Scottish Fishermen's Federation*, which incorporates eight regional and sectoral associations across Scotland. Denmark has essentially three different fishermen's organizations: the *Danish Fishermen's Producer Organisation (Danske Fiskeres Producentorganisation)* contains almost the entire demersal sector and membership exists nation-wide. It is concerned primarily with the implementation of minimum prices and equality of controls. A second demersal Producer's Organisation (*Skagenfishkernes Producentor-*

ganisation) is involved in the organisation of sales contracts. Finally, the *Pelagic Purse Seine Producer Organisation (Notfishkernes Producentorganisation)* has essentially a wide range of responsibilities, which include planning fishing activities for the pelagic sector and the managing herring and mackerel quotas (Symes and Phillipson 1999, 74–76).

7. In his classic sociological study, Weber distinguished the following characteristics of any bureaucracy: hierarchy, impersonality, continuity, and expertise (cf. Weber 1964, 162). In this thesis informational asymmetry shall be added to continuity and expertise, in order to be able to define this bureaucracy. The other two features, namely, hierarchy and impersonality, are not relevant for the preference formation of this political actor.

8. The codecision procedure introduced by the Treaty on European Union in 1993, gives the EP the power to legislate co-equally with the Council of the EU in issues concerning the Single Market. For a detailed overview of how the codecision procedure works in practice see Art. 251 of the EC Treaty and the *Codecision Guide* edited by the Council of the European Union (1999).

9. The Treaty of Nice amended the system of QMV, which will apply after ratification by the 15 national parliaments. In 2005, after the Eastern enlargement, QMV at the Council will then require a triple majority: the number of weighed voted equalling or exceeding the threshold; a simple majority of the Member states; and a supermajority (62 percent) of the population must be represented.

10. During the negotiations on the Single European Act, the actors' preferences diverged on the choice of a future decision rule: Southern countries and Ireland favoured a simple majority voting; traditional reluctant countries, like Great Britain and Denmark preferred to continue with unanimity rule. The French-German tandem and the Benelux countries supported a third alternative: qualified majority voting, which was also adopted for the internal market legislation (König and Bräuninger 1998, 127).

11. Historical institutionalist approaches concentrate on the lags between decisions and long-term consequences, that is the constraints which have appeared over time on EU actors, and particularly the autonomous actions of national government representatives governments as a result of their EU-membership (Pierson 1996).

12. Thanks to Michael Bolle for suggesting this line of argumentation.

13. Spatial models break the analysis of politics into three components: voter choice, party platform selection, and quality of outcomes. In this work the focus shall be on the first component, voters' choice, here translated as member states' most preferred point. For a general overview of spatial competition as a model of political choice see Hinich and Munger (1997).

CHAPTER 3: ACTORS' PREFERENCES AND THE INSTITUTIONAL SETTING IN ACTION

1. Marathon meetings were first adopted in the 1960s in order to deal with difficult issues, like the sites of institutions and the settlement of the common agricultural policy (Hayes-Renshaw and Wallace 1997, 61).

2. Policy entrepreneurs are necessary as agenda-setters for coupling the following three streams: the recognition of a problem, the development of policy proposals, and a receptive political climate. In general, policy entrepreneurs bring new issues into the agenda by providing information and by pushing for specific problem definitions (Kingdon 1984, 19).

3. For a general overview of how a coalition can be negotiated see Bottom et al. 1998.

4. This is also confirmed in the Council of the European Communities (1983) *Minute of the Fisheries Council Meeting* from 03.10.1983, Bobbin No. 3310 p. 7.

5. In the classical sense of coalition formation among political parties, in the term minimal winning coalitions *"winning"* refers to the coalition having the smallest number of votes which nevertheless constitute the majority of parliamentary seats, while *"minimal"* refers to the fact that the cabinet only includes the party or parties necessary to reach a majority in parliament (Lijphart 1999, 90; Kirchsteiger and Puppe 1997, 296).

6. The CAP negotiations during 1984 constitute another prominent example of issue linkage in the EU. At that time, member states became concerned about the increasing proportion of the Community's resources assigned to the European Agricultural Guidance and Guarantee Fund. Those countries with a small agricultural sector and net imports of products covered by CAP regimes were net contributors to the budget. The bargaining position of Great Britain was that this situation was unacceptable due to its relatively high contributions to the EC budget and comparatively small net gains. Thus at the European Council meeting in Fontainebleau in June 1984, the question of how to finance the gap between what was available in the budget was linked to budgetary discipline in the CAP (cf. Weber and Wiesmeth 1991, 256).

7. After a referendum held in September 1972, Norway, decided not to adhere. At that time, fisheries was one of the main obstacles to the accession of Norway, because the fisheries sector feared the consequences of the equal access principle in the fisheries sector, i.e. that a big part of the resources in the Norwegian EEZ would be allocated to other nations which had been fishing there prior to the introduction of the 200 mile zone (Ritchie and Zito 1998, 159). For a general overview of the Norwegian negotiating position on the fisheries issue with the EC see Leigh (1983).

CHAPTER 4: ACTORS' PREFERENCES AND THE INSTITUTIONAL SETTING IN ACTION

1. The central issue for the Nash bargaining solution is to specify the conditions under which successful bargaining between equal strategic actors might be possible. Accordingly, outcomes of bargaining situations have to meet the following four conditions: joint efficiency, symmetry, linear invariance, and independence of irrelevant alternatives. Joint efficiency means that bargaining outcomes must be located on the utility-frontier, so that each party obtains the same benefits. Symmetry, refers to the fact that if the two bargaining parties have the same utility function, they will divide the difference between their reservation points equally. Linear invariance reflects a situation where the utility functions of players are invariable. Finally, independence of irrelevant alternatives implies that the bargaining outcome should not be changed by deleting alternatives or adding new ones that are inferior to the outcome (see Morrow 1994, 113–114).

    The Nash bargaining model, however, is a static model and with some limitations, because the emphasis is placed on the specification of the conditions which characterise bargaining situations. One problem is that the structure of the bargaining game is not taken into consideration. In the Nash bargaining solution players are treated as equals. Another problem is whether its axioms are reasonable. The Nash bargaining solution represents a prediction of the bargaining outcomes that might be reached when the four conditions are fulfilled, but it does not take into account that bargaining structures may lead to asymmetries between the players. The condition of the independence of irrelevant alternatives may be too strong. Furthermore, the Nash bargaining solution does not consider two determinative components of negotiation analysis, namely the interaction of toughness and softness of an actor and the use of threats to move veto players from their bargaining positions. The dilemma of the toughness of a negotiator states that the tougher an actor behaves, the more likely it is that s/he will obtain a larger part of the outcome, but the less likely that any agreement will be reached. On the other hand, the softer an actor behaves, the more likely an agreement will be reached, but the less likely s/he will gain a large part of the outcome. In this situation one might imagine that there is an equilibrium point obtainable through a mixed strategy. Nonetheless, although there have been repeated efforts from different disciplines to settle a normative modelling for negotiations, this dilemma has not yet been resolved (Kremenyuk et al. 2000, 23).

2. The renegotiation of the terms of accession concerned basically Britain's contribution to the Community budget, the length of the transitional period, the operation of the CAP, the retention of national powers in regional, industrial and fiscal policy, among others (Charlot and Sergeant 1986, 61).

3. Because the Labour Party in 1974 was split on the issue whether Britain should remain in Europe, promised during the electoral campaign to hold a referendum on whether to accept or reject the terms of EC-membership. The referendum was held on 5 June 1974 and its result was clearly in favour of remaining: 67 percent of the votes were cast in favour and 33 percent were against continued membership (Charlot and Sergeant 1986, 58, 66).

4. The EC represents only the negotiating position of seven member states, because the British government boycotted the informal meeting on January 1978, in which agreement was reached.

5. France would get 13.6 percent, this being close to its share of 13.8 percent, but less than its maximum demand of 20 percent; Denmark would receive 24.1 percent, approximately equal to its traditional share. For the other member states allocations in line with their traditional shares were also provided, with the exception of Ireland, which progressed from 1.5 percent to 4 percent, in accordance with concessions (side-payments) concluded to buy its acceptance of the conservation and management policy (Leigh 1983, 84, 92–93).

6. The intent was, however, to abolish both boxes after a period of ten years, but this has not yet been achieved.

7. A compromise deal was settled: Great Britain assured continued access the other member state's historic rights to its six and twelve mile zone and to the Shetland and Orkney Boxes, and proposed a symbolic reduction in these rights if they had not been put to use. In return, France would give up its claim of unfettered equal access after December 31, 1982, and would rather accept a formal codification of those rights to be maintained until 2003. The French government could accept this solution, as continued access to the six and twelve zone around the British coast was assured and Britain could not threat to expel French vessels from its waters (Leigh 1983, 86).

8. At the beginning of the 1980s it contributed to 0.7 percent of the Danish national income, while the Community average was of 0.22 percent (Leigh 1983, 96).

9. In 1980, the volume of Denmark's catches amounted to 40 percent of the EC total, followed by Great Britain with 16 percent, France 15 percent, and Italy 8 percent. Thereby, more than two thirds of catch) consists of small pelagic species destined for fish meal (Salz 1991, 79). Since the accession of Spain and Portugal, Spain became the most important fish nation in terms of catches. Spanish high value species for human consumption are worth more than five

times as much as the Danish catch, as the latter consist of "industrial species" such as sandeel, sprat and Norway pount which will be reduced to fish meal and oil. For example, a tonne of sandeel just brings a fraction of what would be paid for a tonne of cod (Leigh 1983, 12).

10. Since Danish vessels had not traditionally fished in that area, Denmark was not among the member states to receive licences to fish for edible demersal species.

11. The so-called Luxembourg compromise was agreed upon during the latter half of 1965. It stated essentially that the Council would not vote on a proposal, if a member state declared that to do so would affect its "vital national interests," which did not have to be specified.

12. This is clearly stated in the Council of the European Communities (1983) *Minute of the Fisheries Council Meeting* from 25.01.1983, Bobbin No. 2594, p. 5–6.

13. I thank Thomas König for drawing my attention to this issue.

14. From the beginning of EC membership, the Parliament and the political parties played a central role in Denmark's policy towards the EU, because since 1973 Denmark has been governed by minority governments with weak parliamentary basis. They had to take the other parties into account and permanently built *ad hoc* coalitions on every issue. On average, every new government stayed two years in power (Green-Pedersen 2001, 10–11).

15. This was also clearly stated in the Council of the European Communities (1983) *Minutes of the Fisheries Council* from 25.01.1983, Bobbin No. 2595, p. 8.

16. It goes beyond the scope of this study to analyse in detail the role played by the European Court of Justice on fisheries conservation and management policy. For a detailed study on its role in the European integration process in general see Alter (1998).

17. This is confirmed in the internal discussion paper published by the European Commission 1992, 24.

18. "Within the passing years the CFP framework has been built up piece by piece. To change it would be terrible. We did not extend the cake; we have just dusted it and we have tied it up like a package and its presentation is completed. There is always a moment when a bargaining process has to be completed and I think that this moment has come. These are crucial moments and if we prolong the discussion we are going to ruin it and shall to have to start from scratch. This would be extremely dangerous." (translation by the author). As quoted in the stenographic record of the Council of the European Communities (1983) *Minutes of the Fisheries Council* from 25.01.1983, Bobbin No. 2594, p. 4.

19. Hirschmann conceived exit and voice not only as an alternative mechanism for the individual's reaction to the performance of institutions of which s/he is member, but also established a negative association between the two: the opportunities for exit reduce the need to voice. The lack of exit may enhance the willingness for the voice (Bartolini 1998, 10; Pollack 2001, 223).

20. This is clearly stated in the following internal Council document: Council of the European Communities *Projet du Procès Verbal* from 18–19 December 1989, p. 15–17.

21. Nevertheless, the Portuguese and the Spanish fleets are still not allowed to fish in the North Sea, which contains some 90 percent of all EU fish stocks.

22. Kandogan (2001) comes to similar conclusions. Using a game theoretic model of decision making and drawing upon the experience of previous enlargements, he argued that pre-accession reforms to change budgetary rules to limit the structural funds that will flow to the East, will be ineffective in reducing budgetary outlays, because Eastern European countries will reverse the reformed rules by forming coalitions among themselves and with some of the incumbent countries with similar interests.

Bibliography

Abélès, Marc, and Irène Bellier. "La Commission européenne: du compromis culturel à la culture politique du compromis. " *Revue Française de Science Politique* 46 (1996): 431–456.

Alter, Karen J. "Who are the 'Masters of the Treaty'? European Governments and the European Court of Justice." *International Organization* 52 (1998): 121–148.

Anderson, Lee G. "Privatizing Open Access Fisheries: Individual Transferable Quotas." In *Handbook of Environmental Economics*, edited by D. W. Bromley. Oxford: Blackwell Publishers Ltd., 1996.

Armstrong, Kenneth, and Simon J. Bulmer. *The Governance of the Single European Market*. Manchester: Manchester University Press, 1998.

Arregui, Javier. "Bargaining and Decision-Making in the European Union. Description and Application of Two Alternative Bargaining Models to the European Directive on the Manufacture, Presentation and Sale of Tobacco Products." Paper prepared for presentation at the 4th Pan-European International Relations Conference at the University of Kent Canterbury, 8–10 September 2001.

Aspinwall, Mark. "Structuring Europe: Powersharing Institutions and British Preferences on European Integration." *Political Studies* 48 (2000): 415–442.

Aspinwall, Mark, and Gerald Schneider. "Same menu, separate tables: The institutionalist turn in political science and the study of European integration." *European Journal of Political Research* 38 (2000): 1–36.

Ávila de Melo, António, and Ricardo de Alpoim. *Recursos Pesqueiros do Noroeste Atlântico: Situação actual com base nas avaliações de 1998 feitas pelo Concelho Científico da NAFO*. Lisboa: Instituto de Investigação das Pescas e do Mar, 1999.

Axelrod, Robert. *The Evolution of Cooperation*. New York: Basic Books, 1984.

Baron, David P., and John A. Ferejohn. "Bargaining in legislatures." *American Political Science Review* 83 (1989): 1181–1206.

Bartolini, Stefano. "Exit Options, Boundary Building, Political Structuring." Florence: European University Institute Working Paper SPS 98/1, 1998.

Berejekian, Jeffrey. "The Gains Debate: Framing State Choice." *American Political Science Review* 91 (1997): 789–805.

Beyers, Jan, and Guido Dierickx. "The Working Groups of the Council of the European Union: Supranational or Intergovernmental Negotiations?" *Journal of Common Market Studies* 36 (1998): 289–317.

Beyme, Klaus von. *Interessengruppen in der Demokratie*. 5. überarbeitete Neuauslage. München : Piper Verlag, 1980.

Binmore, Ken. "Game Theory and the Social Contract.". In *Game Equilibrium Models II. Methods, Morals, and Markets* edited by R. Selten. Berlin: Springer Verlag, 1991.

Bottom, William P., et al., "Negotiating a Coalition: Risk, Quota Shaving, and Learning to Bargain." Paper prepared for the Conference on Experimental Economics: Research on Learning and Bargaining held at Washington University in May 1998.

Bromley, David W., and Jeffrey A. Cochrane. "A Bargaining Framework for the Global Commons." In *Handbook of Environmental Economics*, edited by D. W. Bromley. Oxford: Blackwell Publishers Ltd., 1996.

Buchanan, James. "Social Choice, Democracy, and Free Markets." *Journal of Political Economy* LXII (1954): 114–123.

–––– *The Limits of Liberty. Between Anarchy and Leviathan.* Chicago: University of Chicago Press, 1975.

Buchanan, James M., and Gordon Tullock. *The Calculus of Consent. Logical Foundation of Constitutional Democracy.* Michigan: Ann Arbor, 1962.

Bueno de Mesquita, Bruce. "Political Forecasting: An Expected Utility Method." In *European Community Decision Making. Models, Applications, and Comparisons*, edited by B. Bueno de Mesquita, and F. N. Stokman. New Haven, London: Yale University Press, 1994.

––––"A Decision Making Model: Its Structure and Form." *International Interactions* 23 (1997): 235–266.

–––– *Principles of International Politics. People's Power, Preferences, and Perceptions.* Washington D.C.: Congressional Quarterly, 2000.

Bueno de Mesquita, Bruce, and Frans N. Stokman. *European Community Decision Making. Models, Applications, and Comparisons.* New Haven: Yale University, 1994.

Bulmer, Simon. "The Governance of the European Union: A New Institutionalist Approach." *Journal of Public Policy* 13 (1994): 351–380.

–––– "New Institutionalism and the governance of the Single European Market." *Journal of European Public* Policy 5 (1998): 365–386.

Cann, Charles. *Saving Our Fish.* London: Centre for European Reform, 1997.

Charlot, Monica, and Jean-Claude Sergeant. *Britain and Europe since 1945.* Paris: Colin-Longmann, 1986.

Christiansen, Thomas et al. "The social construction of Europe." *Journal of European Public Policy* 6 (1999): 528–544.

Coffey, Clare. *Introduction to the Common Fisheries Policy: An Environmental Perspective.* London: Institute for European Environmental Policy, 1995.

–––– "European Funding for Sustainable Development of Fisheries Dependent Regions." In *Fisheries Dependent Regions*, edited by D. Symes. Oxford: Fishing News Books, 2000.

Colomer, Joseph M., and Madeleine Hosli. "Decision-Making in the European Union: The Power of Political Parties." In *Decision Rules in the European Union: A Rational Choice Perspective* edited by P. Moser et al. Basingstoke: MacMillan Press Ltd., 2000.

Commission of the European Communities. *Report by the Commission to the Council and Parliament on the Application of the Act of Accession of Spain and Portugal in the Fisheries Sector.* Brussels: SEC (92) 2340 final, 1992.

Conceição-Heldt, Eugénia da. *Dezentralisierungstendenzen in westeuropäischen Ländern. Territorialreformen Belgiens, Spaniens und Italiens im Vergleich.* Berlin: Arno Spitz Verlag, 1998.

Conconi, Paula, and Carlo Perroni. "Issue Linkage and issue tie-in in multilateral negotiations." *Journal of International Economics* 57 (2001): 423–447.

Conrad, Jon M. "Bioeconomic Models of the Fishery." In *Handbook of Environmental Economics*, edited by D. W. Bromley. Oxford: Blackwell Publishers Ltd., 1996.

Council of the European Communities. *Minute of the Fisheries Council Meeting* from 25.01.1983, Bobbin No. 2594 to 2596. Brussels, 1983.

—— *Minute of the Fisheries Council Meeting* from 03.10.1983, Bobbin No. 3310. Brussels, 1983.

—— *Projet de Procès-Verbal de la 1380 èmesession du Conseil Pêche tenue à Bruxelles, les lundi 18 et mardi 19 décembre 1989*, Internal Document. No. 11089/89. Brussels, 1989.

Council of the European Union *Council Guide IV. Codecision Guide*. Internal Document SN 3631/99. Brussels, 1999.

Cram, Laura. "The European Commission as a Multi-Organization: Social Policy and IT Policy in the EU." *Journal of European Public Policy* 1(1994): 195–217.

—— "Whither the Commission? Reform, renewal and the issue-attention cycle." *Journal of European Public Policy* 8 (2001): 770–786.

Cram, Laura et al. "Reconciling Theory and Practice." In *Developments in the European Union*, edited by L. Cram et al. Basingstoke: MacMillan Press Ltd., 1999.

Cross, John G. "Negotiation as Adaptive learning." *International Negotiation* 1 (1996): 153–178.

Cushing, David. *Population production and regulation in the sea. A fisheries perspective*. Cambridge: Cambridge University Press, 1995.

Dogan, Rhys. "A Cross-sectoral View of Comitology: Incidence, Issues and Implications." In *Europe in Change: Committee Governance in the European Union*, edited by T. Christiansen, and E. Kirchner. Manchester: Manchester University Press, 2000.

Doleys, Thomas J. "Member states and the European Commission: theoretical insights from the new economics of organization." *Journal of European Public Policy* 4 (2000): 532–553.

Downs, Anthony. *An Economic Theory of Democracy*. New York: Harper & Brothers, 1957.

Dupont, Christophe. "Negotiation as Coalition Building." *International Negotiation* 1 (1996): 47–64.

Dupont, Christoph, and Guy-Olivier Faure. "The Negotiation Process." In *International Negotiation. Analysis, Approaches, Issues*, edited by V. A. Kremenyuk. San Francisco: Jossey-Bass Publishers, 1991.

Eichener, Volker. *Das Entscheidungssystem der Europäischen Union. Institutionelle Analyse und demokratietheoretische Bewertung*. Opladen: Leske + Budrich, 2000.

Elgström, Olen. " 'The honest broker'?—The EU Council Presidency as a mediator." Paper presented at the 4th Pan-European International Relations Conference at the University of Kent Canterbury, 8–10 September 2001.

Elgström, Olen, and Christer Jönsson. "Negotiation in the European Union: bargaining or problem-solving?." *Journal of European Public Policy* 7 (2000): 684–704.

Elgström, Olen, and Michael Smith. "Introduction: Negotiation and policy-making in the European Union — processes, system and order." *Journal of European Public Policy* 7 (2000): 673–683.

Eriksen, Erik O. "The Question of Deliberative Supranationalism in the EU." Oslo: Arena Working Paper 99/4, 1999.

European Commission. *Report by the Commission to the Council and Parliament on the Application of the Act of Accession of Spain and Portugal in the Fisheries*

Sector. Luxembourg: Office for Official Publications of the European Communities, 1992.

European Commission. *Green Paper on the future of the common fisheries policy. Volume I.* Luxembourg: Office for Official Publications of the European Communities, 2001.

Farnell, John, and James Elles. *In Search of a Common Fisheries Policy.* Aldershot: Gower Publishing Company, 1984.

Fearon, James D. "Bargaining, Enforcement, and International Cooperation." *International Organization* 52 (1998): 269–305.

Feyerabend, Paul K. *Probleme des Empirismus. Schriften zur Theorie der Erklärung, der Quantentheorie und der Wissenschaftsgeschichte.* Braunschweig: Vieweg, 1981.

Flaaten, Ola, *et al.,* "Fisheries Management under uncertainty—an overview." *Fisheries Research* 37 (1998): 1–6.

Fligstein, Neil, and Jason McNichol. "The Institutional Terrain of the European Union." In *European Integration and Supranational Governance,* edited by W. Sandholtz, and A. Stone Sweet. Oxford: Oxford University Press, 1998.

Foders, Federico. *Reforming the European Union's Common Fisheries Policy. Issues in Conservation and Policy Option.* London: European Policy Forum, 1994.

Food and Agriculture Organization of the United Nations (FAO) *The State of the World Fisheries and Aquaculture.* Rome: FAO, 1997.

Franchino, Fabio. "Control of the Commission's Executive Functions: Uncertainty, Conflict and Decision Rules." *European Union Politics* 1 (2000a): 59–88.

———"Commission's Executive Discretion, Information and Comitology." *Journal of Theoretical Politics* 12 (2000b): 155–181.

Franke, Siegfried F. "Opportunismus als Form politischer Rationalität der öffentlichen Verwaltung." *Jahrbuch für Neue Politische Ökonomie* 8 (1989): 158–171.

Freire, Juan, and Antonio García-Allut. "Socioeconomic and biological causes of management failures in European artisanal fisheries: the case of Galicia (NW Spain)." *Marine Policy* 24 (2000): 375–384.

Friis, Lykke. *When Europe Negotiates: From Europe Agreements to Eastern Enlargement.* Copenhagen: Ph.D. Dissertation Institute of Political Science, University of Copenhagen 1997.

Friis, Lykke, and Anna Murphy. "The European Union and central and Eastern Europe: Governance and Boundaries." *Journal of Common Market Studies* 37 (1999): 211–232.

Gamson, William A. "A Theory of Coalition Formation." *American Sociological Review* 26 (1961): 373–382.

Garcia, Serge M., and Masaki Hayashi. "Division of the oceans and ecosystem management: A contrastive spatial evolution of marine fisheries governance." *Ocean and Coastal Management* 43 (2000): 445–474.

Garrett, Geoffrey, and George Tsebelis. "An institutional critique of intergovernmentalism." *International Organization* 50 (1996): 269–299.

Garza-Gil, Dolores, et al. "The Spanish case regarding fishing regulation." *Marine Policy* 20 (1996): 249–259.

Gilardi, Fabrizio. "Principal-agent models go to Europe: independent regulatory agencies as ultimate step of delegation." Paper presented at the ECPR General Conference at the University of Kent Canterbury, 6–8 September 2001.

González Laxe, Fernando. "The Inadequacies and Ambiguities of the Common Fisheries Policy." In *Alternative Management Systems for Fisheries,* edited by D. Symes. Oxford: Fishing News Books, 1999.

Grande, Edgar. "Multi-Level Governance: Institutionelle Besonderheiten und Funktionsbedingungen des europäischen Mehrebenensystems." In *Wie Problemlösungsfähig ist die EU? Regieren im europäischen Mehrebenensystem*, edited by E. Grande, and M. Jachtenfuchs. Baden-Baden: Nomos Verlag, 2000.

Gray, Pat. "Policy Disasters in Europe: An Introduction." In *Public Policy Disasters in Western Europe*, edited by P. Gray, and P.'t Hart. London and New York: Routledge, 1998.

Gray, Tim S. *The Politics of Fishing*. Hampshire: MacMillan Press Ltd., 1998.

Green-Pedersen, Christoffer. "Minority Governments and Party Politics: The Political and Institutional Background to the 'Danish Miracle'". Köln, Max-Planck-Institut für Gesellschaftsforschung Discussion Paper 01/1, 2001.

Grieco, Joseph M. "State Interests and Institutional Rule Trajectories: A Neorealist Reinterpretation of the Maastricht Treaty and European Economic and Monetary Union." In *Realism: Restatements and Renewal*, edited by B. Frankel. London: Frank Cass, 1996.

Haas, Ernst B. "International Integration. The European and the Universal Process." *International Organization* XV (1961): 366–392.

Hall, Peter A., and Rosemary C.R. Taylor. "Political Science and the Three New Institutionalismus." *Political Studies* XLVI (1996): 958–962.

Hanna, Susan. "Parallel Institutional Pathologies in Fisheries Management." In *Northern Waters: Management Issues and Practice*, edited by D. Symes. Oxford: Fishing News Books, 1998.

Hardin, Garrett. "The Tragedy of the Commons." *Science* 162 (1968): 1243–1248.

Harnier, Otto. "Gemeinsame Fischereipolitik." In *Handbuch der europäischen Integration*, edited by M. Röttinger, and C. Weyringer. Wien: Manz, 1996.

Hatcher, Aaron C. "Subsidies for European fishing fleets: the European Community's structural policy for fisheries 1971–1999." *Marine Policy* 24 (2000): 129–140.

Hayes-Renshaw, Fiona, and Helen Wallace. "Executive power in the European Union: the functions and limits of the Council of Ministers." *Journal of European Public Policy* 2 (1995): 559–582.

—— *The Council of Ministers*. Hampshire and London: Macmillan Press Ltd., 1997.

Héritier, Adrienne. "The Accommodation of Diversity in European Policy Making and its Outcomes: Regulatory Policy as a Patchwork." Florence: European University Institute Working Paper, SPS 96/2, 1996.

Hinich, Melvin J. and Michael C. Munger. *Analytical Politics*. Cambridge: Cambridge University Press, 1997.

Hirschman, Albert O. *Exit, voice, and loyalty: responses to decline in firms, organizations and states*. Cambridge, MA: Harvard University, 1975.

Hix, Simon, and Klaus H. Goetz. "Introduction: European Integration and National Political Systems." *West European Politics* 23 (2000): 1–26.

Hoffman, Stanley. "Obstinate or obsolete? The Fate of the Nation-State and the Case of Western Europe." *Daedalus* 95 (1966): 862–915.

——"Reflections on the Nation-State in Western Europe Today." *Journal of Common Market Studies* 21 (1982): 21–37.

Holden, Mike. *The Common Fisheries Policy. Origin, Evaluation and Future*. Oxford: Fishing News Books, 1996.

Honneland, Geir. "A model of compliance in fisheries: theoretical foundations and practical application." *Ocean and Coastal Management* 42 (1999): 699–716.

Hooghe, Liesbet. *The European Commission and the Integration of Europe*. Cambridge: Cambridge University Press, 2001.

Hosli, Madeleine O. "The Political Economy of Subsidiarity." In *The Political Economy of the European Integration*, edited by F. Laursen. The Hague: Kluwer Law International, 1995.

——"Coalitions and Power: Effects of Qualified Majority Voting in the Council of the European Union." *Journal of Common Market Studies* 34 (1996): 255–273.

Hug, Simon, and Thomas König. "In View of Ratification: Governmental Preferences and Domestic Constraints at the Amsterdam Intergovernmental Conference." *International Organization* 56 (2002): 447–476.

Iida, Keisuke. "When and How Do Domestic Constraints Matter? Two-level Games with Uncertainty." *Journal of Conflict Resolution* 37 (1993): 403–426.

Institut Français pour l'exploitation de la mer (IFREMER). *Evaluation of the Fisheries Agreements concluded by the European Community.* Issy-les-Moulineaux: IFREMER, 1999.

Jönsson, Christer. "Conceptualizations of the Negotiation Process." Paper prepared for presentation at the 4th Pan-European International Relations Conference at the University of Kent Canterbury, 8–10 September 2001.

Jupille, Joseph, and James A. Caporaso. "Institutionalism and the European Union: Beyond International Relations and Comparative Politics." *Annual Review of Political Science,* 2 (1999): 429–444.

Kandogan, Yener. "Political economy of eastern enlargement of the European Union: Budgetary costs and reforms in voting rules." *European Journal of Political Economy* 16 (2001): 685–705.

Karagiannakos, Apostolos. *Fisheries Management in the European Union.* Avebury: Ashgate Publishing, 1995.

——"Total Allowable Catch (TAC) and quota management system in the European Union." *Marine Policy* 20 (1996): 235–248.

Kato, Junko. "Review Article: Institutions and Rationality in Politics –Three Varieties of Neo-Institutionalists." *British Journal of Political Science* 26 (1996): 553–582.

Keohane, Robert, and Joseph S. Nye. *Power and Interdependence.* Boston: Little & Brown, 1977.

Keohane, Robert O., and Stanley Hoffman. "Conclusions: Community Politics and Institutional Change." In *The Dynamics of the European Integration,* edited by W. Wallace. London: Pinter, 1990.

Kingdon, John W. *Agendas, Alternatives, and Public Policies.* Boston: Little & Brown, 1984.

Kirchsteiger, Georg, and Clemens Puppe. "On the Formation of Political Coalitions." *Journal of Institutional and Theoretical Economics* 153 (1997): 293–319.

Kohler-Koch, Beate. "Catching up with change: the transformation of governance in the European Union." *Journal of European Public Policy* 3 (1996): 359–380.

Koelble, Thomas A. "The New Institutionalism in Political Science and Sociology." *Comparative Politics* 27 (1995): 1–39.

König, Thomas and Thomas Bräuninger. "The Inclusiveness of European Decision Rules." *Journal of Theoretical Politics* 10 (1998): 125–142.

——"The Eastern Enlargement of the European Union. Accession Scenarios, Institutional Reform and Policy Positions." Paper presented at the Deutsche Forschungsgemeinschaft Conference , 1.–2. November 2001, Mannheim.

König, Thomas, and Mirja Pöter. "Exploring the domestic Arena. The Formation of Policy Positions in EU Member States." Paper presented at the 4th Pan-European International Relations Conference at the University of Kent Canterbury, 8–10 September 2001.

Koremenos, Barbara et al., "Rational Design: looking Back to Move Forward." *International Organization* 55 (2001): 1051–1082.

Kremenyuk, Victor et al., "International Economic Negotiation: Research Tasks and Approaches." In *International Economic Negotiation: Models versus Reality,* edited by V. Kremenyuk, and G. Sjöstedt. Cheltenham: Edward Elgar, 2000.

Laruelle, Anick, and Mika Widgrén. "Is the Allocation of voting power among EU states fair?" *Public Choice* 94 (1998): 317–339.

Lacy, Dean, and Emerson M.S. Niou. "Nonseparable Preferences, Issue Linkage, and economic sanctions." Paper prepared for presentation at the Annual Meeting of the American Political Science Association, Boston, Massachusetts, 3–6 September 1998.

Lakatos, Imre. "Falsification and the methodology of scientific research programmes." In *Criticism and the Growth of Knowledge*, edited by I. Lakatos, and A. Musgrave. Cambridge: Cambridge University Press, 1970.

——"Lectures on Scientific Method." In *For and Against Method: including Lakato's lectures on scientific method and the Lakatos-Feyerabend correspondence*, edited by M. Motterlini. Chicago: Chicago University Press, 1999.

Lasswell, Harold. *Politics: who gets what, when, how*. Eleventh printing. New York: World Publishing Company, 1971.

Laver, Michael, and Norman Schofield. *Multiparty Government. The Politics of Coalition in Europe*. Oxford: Oxford University Press, 1990.

Lax, David A., and James K. Sebenius. *The Manager as a Negotiator*. New York: Free Press, 1986.

Leigh, Mike. *European Integration and the Common Fisheries Policy*. London: Croom Helm, 1983.

Lequesne, Christian. "La Commission européenne entre autonomie et dépendance." *Revue Française de Science Politique* 46 (1996): 389–408.

——"Quand l'Union européenne gouverne les poissons: pourquoi une politique commune de la pêche?." Les Études du Centre d'études et de recherches internationales Sciences Po (CERI) No. 61. Paris: CERI, 1999.

——"The Common Fisheries Policy. Letting the Little Ones Go?." In *Policy-Making in the European Union*, edited by H. Wallace, and W. Wallace. 4th ed. Oxford: Oxford University Press, 2000a.

——"Quota Hopping: The Common Fisheries Policy Between States and Markets." *Journal of Common Market Studies* 38 (2000b): 779–793.

—— *L'Europe Bleue. À quoi sert une politique communautaire de la pêche?.* Paris: Presses de Sciences Po, 2001.

Lewis, Jeffrey. "The Institutional Problem-Solving Capacities of the Council: The Committee of Permanent Representatives and the Methods of Community." Cologne: Max-Planck-Institut für Gesellschaftsforschung Discussion Paper 98/1, 1998.

Lijphart, Arend. *Patterns of Democracy. Government Forms and Performance in Thirty-Six Countries*. New Haven: Yale University Press, 1999.

Lindberg, Leon N., and Stuart Scheingold. *Europe's Would-Be Polity. Patterns of Change in the European Community*. New Jersey: Prentice-Hall, Inc., 1970.

Lodge, Juliet E., and Frank R. Pfetsch. "Negotiating the European Union: Introduction." *International Negotiation* 3 (1998): 289–292.

Mackinson, Steven, et al., "Bioeconomics and catchability: fish and fishermen behaviour during stock collapse." *Fisheries Research* 31 (1997): 11–17.

Majone, Giandomenico. "The Rise of the Regulatory State in Europe.". *West European Politics* 17 (1994): 77–101.

——"Two Logics of Delegation. Agency and Fiduciary Relations in EU Governance.". *European Union Politics* 2 (2001): 103–122.

March, James C., and Johan P. Olsen. *Rediscovering Institutions*. New York: Free Press, 1989.

Marks, Gary, et al., "Integration Theory, Subsidiarity and the Internationalisation of Issues: The Implication for Legitimacy." Florence: European University Institute Working Paper, RSC No. 95/7, 1995.

Mayer, Frederick W. "Managing domestic differences in international negotiations: the strategic use of internal side-payments." *International Organization* 46 (1992): 793–818.

McCubbins, Matthew D., and Thomas Schwartz. "Congressional Oversight Overlooked: Police Patrols versus Fire Alarms." *American Journal of Political Science* 28 (1984): 165–179.

McGinley, Joan. *Ireland's Fishery Policy*. Donegal: Croaghlin, 1991.

Milgrom, Paul, and John Roberts. *Economics, Organization and Management*. Englewood Cliffs: Prentice Hall, 1992.

Milner, Helen V. *Interests, institutions, and information: Domestic politics and international relations*. Princeton: Princeton University Press, 1997.

Milner, Helen, and B. Peter Rosendorff. "Democratic Politics and International Trade Negotiations. Elections and Divided Government as Constraints on Trade Liberalization." *Journal of Conflict Resolution* 41 (1997): 117–146.

Mitchell, Carlyle L. "Fisheries management in the Grand Banks, 1980–1992 and the straddling stock issue." *Marine Policy* 21 (1997): 97–109.

Mo, Jongryn. "The Logic of Two-Level Games with Endogenous Domestic Coalitions." *Journal of Conflict Resolution* 38 (1994): 402–422.

—— "Domestic Institutions and International Bargaining: The Role of Agent veto in Two-Level Games." *American Political Science Review* 89 (1995): 914–924.

Moravcsik, Andrew. "Negotiating the Single European Act: national interests and conventional statecraft in the European Community." *International Organization* 45 (1991): 19–56.

—— "Preferences and Power in the European Community: A Liberal Intergovernmentalist Approach." *Journal of Common Market Studies* 31 (1993): 473–524.

—— "Taking Preferences Seriously: A Liberal Theory of International Politics." *International Organization* 51 (1997a): 513–553.

—— "Does the European Union represent an *n* of 1?." *ECSA Review* X (1997b): 4–8.

—— *The Choice for Europe. Social Purpose and State Power from Messina to Maastricht*. Ithaca, NY: Cornell University Press, 1998.

Moravcsik, Andrew, and Kalypso Nicolaidis. "Explaining the Treaty of Amsterdam: Interests, Influence, Institutions." *Journal of Common Market Studies* 37 (1999): 59–85.

Morin, Michel. "The fisheries resources in the European Union. The distribution of TACs: principle of relative stability and quota-hopping." *Marine Policy* 24 (2000): 375–384.

Morrow, James D., *Game Theory for Political Scientists*. Princeton: Princeton University, 1994.

Mueller, Dennis C., *Public Choice II*. Rev. ed. Cambridge, MA: Cambridge University Press, 1995.

Nannestad, Peter. "Das Politische System Dänemarks." In *Die Politischen Systeme Westeuropas*, edited by W. Ismayr. Opladen: Leske + Budrich, 1997.

Nash, John. "The Bargaining Problem." *Econometrica* 18 (1950): 155–162.

Niskanen, William A. *Bureaucracy and Representative Government*. Chicago: Aldine-Atherton, 1971.

Nugent, Neill. "The Leadership Capacity of the European Commission." *Journal of European Public Policy* 2 (1995): 603–623.

—— *The Government and Politics of the European Union*. 3rd Edition. Basingstoke: Macmillan, 1999.

Oakerson, Ronald J. "Analyzing the Commons: A Framework." In *Making the Commons Work. Theory, practice, and Policy*, edited by D.W. Bromley. San Francisco: Institute for Contemporary Studies, 1992.

Organisation de Coopération et de Dévelopement Économiques (OECD). *Examen des Pêcheries dans les Pays de l'OCDE*. Paris: OCDE, 1996.

Ostrom, Elinor. "New Horizons in Institutional Analysis." *American Political Science Review* 89 (1995): 174–178.

Pahre, Robert. "Endogenous Domestic Institutions in Two-Level Games and Parliamentary Oversight of the European Union." *Journal of Conflict Resolution* 41 (1997): 147–174.

Patterson, Lee A. "Agricultural policy reform in the European Community: a three-level game analysis." *International Organization* 51 (1997): 135–165.

Payne, Dexter C. "Policy-Making in Nested Institutions: Explaining the Conservation Failure of the EU's Common Fisheries Policy." *Journal of Common Market Studies* 38 (2000): 303–324.

Peters, Guy B. "Bureaucratic Politics and the Institutions of the European Community." In *Euro-Politics. Institutions and Policy-Making in the "New" European Community*, edited by A. Sbragia. Washington, D.C.: Brookings Institution, 1992.

———"Institutional Theory: Problems and Prospects." Wien: Institut für Höhere Studien, Reihe Politikwissenschaft, 2000.

Peterson, John. "The Choice for EU theorists: Establishing a common framework for analysis." *European Journal of Political Research* 39 (2001): 289–318.

Pfetsch, Frank R. "Negotiating the European Union: A Negotiation-Network Approach." *International Negotiation* 3 (1998): 293–317.

Pierce, Roy. "Introduction: Fresh Perspectives on a Developing Institution." In *European Community Decision Making. Models, Applications, and Comparisons*, edited by B. Bueno de Mesquita, and F. N. Stokman. New Haven: Yale University Press, 1994.

Pierson, Paul. "The Path to European Integration: A Historical Institutionalist Analysis." *Comparative Political Studies* 29 (1996): 123–163.

——— "The Limits of Design: Explaining Institutional Origins and Change." *Governance* 13 (2000): 475–499.

Plott, Charles. "A Notion of Equilibrium and its Possibility Under Majority Rule." *American Economic Review* 57 (1967): 787–806.

Pollack, Mark A. "Creeping Competence: The Expanding Agenda of the European Community." *Journal of Public Policy* 14 (1994): 95–145.

——— "Delegation, agency, and agenda setting in the European Community." *International Organization* 51 (1997): 99–134.

———"International Relations Theory and European Integration." *Journal of Common Market Studies* 39 (2001): 221–244.

Popper, Karl R. *The Myth of the Framework. In defence of science and rationality*. London: Routledge, 1994.

Powell, Robert. "Absolute and Relative Gains in International Theory." *American Political Science Review* 85 (1991): 1303–1320.

Putnam, Robert D. "Diplomacy and Domestic Politics: the logic of two–level games." *International Organization* 42 (1988): 427–460.

Raiffa, Howard. *The Art and Science of Negotiation*. Cambridge: Harvard University Press, 1982.

Rector, Chad. "Buying Treaties with Cigarettes: Internal Side-payments in Two Level Games." *International Interactions* 27 (2001): 207–238.

Rettig, Bruce R. "Management Regimes in Ocean Fisheries." In *Handbook of Environmental Economics*, edited by D. W. Bromley. Oxford: Blackwell Publishers, 1996.

Richardson, Jeremy. "Policy-making in the EU: Interests, ideas and garbage cans of primeval soup." In *European Union. Power and Policy-Making*, edited by J. Richardson. London: Routledge, 1996.

Riker, William. *The Theory of Political Coalitions*. New Haven: Yale University Press, 1962.

Ritchie, Ellen, and Anthony Zito. "The Common Fisheries Policy. A European Disaster?" In *Public Policy Disasters in Western Europe*, edited by P. Gray, and P. Hart. London and New York: Routledge, 1998.

Risse-Kappen, Thomas. "Exploring the Nature of the Beast: International Relations Theory and Comparative Policy Analysis Meet the European Union." *Journal of Common Market Studies* 34 (1996): 53–80.

Risse, Thomas. " 'Let's argue!' Communicative Action in World Politics." *International Organization* 54 (2000): 1–39.

Rubin, Jeffrey Z., and Bert R. Brown. *The Social Psychology of Bargaining and Negotiation*. Orlando: Academic Press, 1975.

Rubinstein, Ariel. "Perfect equilibrium in a bargaining model." *Econometrica* 50 (1982): 97–109.

Salz, Pavel. *The European Atlantic Fisheries. Structure, economic performance and policy*. The Hague: Agricultural Economics Research Institute, 1991.

Sandberg, Per, et al. "Bioeconomic advice on TAC—the state of the art in the Norwegian fishery management." *Fisheries Research* 37 (1998): 259–274.

Scharpf, Fritz W. "The Joint-Decision Trap: Lessons from German Federalism and European Integration." *Public Administration* 1 (1988): 239–278.

——, 1997. *Games Real Actors Play. Actor-Centered Institutionalism in Policy Research*. Boulder: Westview Press.

Schelling, Thomas. *The Strategy of Conflict*. Cambridge: Harvard University Press, 1960.

Schmidt, Susanne K. "Behind the Council Agenda: The Commission's Impact on Decisions." Cologne: Max-Planck-Institut für Gesellschaftsforschung Discussion Paper 97/4, 1997.

Schneider, Gerald, and Lars-Erik Cederman. "The change of tide in political cooperation: A limited information model of European integration." *International Organization* 48 (1994): 633–662.

Schneider, Volker, and Werle Raymund. "Vom Regime zum korporativen Akteur. Zur institutionellen Dynamik der Europäischen Gemeinschaft." In *Regime in den internationalen Beziehungen*, edited by B. Kohler-Koch. Baden-Baden: Nomos Verlag, 1989.

Schofield, Norman. "Generic Instability of Majority Rule." *Review of Economic Studies* 50 (1983): 696–705.

Sebenius, James K. "Negotiation arithmetic: adding and subtracting issues and parties." *International Organization* 37 (1983): 281–316.

—— "Negotiation Analysis." In *International Negotiation. Analysis, Approaches, Issues*, edited by V. A. Kremenyuk. San Francisco: Jossey-Bass Publishers, 1991.

Shackleton, Michael. "Fishing for a Policy? The Common Fisheries Policy of the Community." In *Policy-Making in the European Community*, edited by H. Wallace, et al. Chichester: John Wiley and Sons, 1983.

Shepsle, Kenneth A., and Mark S. Bonchek. *Analyzing Politics. Rationality, Behavior, and Institutions*. New York: W. W. Norton & Company, 1997.

Sherrington, Philippa. *The Council of Ministers. Political Authority in the European Union*. London and New York: Pinter, 2000.

Smyrl, Marc E. "When (and how) Do the Commission's Preferences Matter?." *Journal of Common Market Studies* 36 (1998): 79–99.

Snidal, Duncan. "Relative Gains and the Pattern of International Cooperation." *American Political Science Review* 85 (1991): 701–726.

Sæter, Martin. "Norway and the European Union." In *The European Union and the Nordic Countries*, edited by L. Miles. London: Routledge, 1996.

Solow, Robert. "A Native Informant Speaks." *Journal of Economic Methodology* 8 (2001): 111–112.

Song, Yann-Huei. "The EC's Common Fisheries Policy in the 1990s." *Ocean Development and International Law* 26 (1995): 31–55.

Spence, David. "Negotiations, coalitions and the resolution of inter-state conflicts." In *The Council of the European Union*, edited by M. Westlake. London: Cartermill Publishing, 1995.

Steins, Nathalie A., and Victoria M. Edwards. "Institutional Analysis of UK coastal fisheries: implications of overlapping regulations for fisheries management." *Marine Policy* 21 (1997): 535–544.

Steunenberg, Bernard, and Torsten Selck. "The Insignificance of the Significance." Paper presented at the *4th Pan-European International Relations Conference* at the University of Kent Canterbury, 8–10 September 2001.

Stevenson, Glenn G. *Common Property Economics. A General Theory and Land Use Applications*. Cambridge: Cambridge University Press, 1991.

Symes, David. "Fishing in Troubled Waters." In *Fisheries Management in Crisis*, edited by K. Crean, and D. Symes. Oxford: Fishing News Books, 1996.

—— "The European Community's Common Fisheries Policy." *Ocean and Coastal Management* 35 (1997): 137–155.

—— *Fisheries Dependent Regions*. Oxford: Fishing News Books, 2000.

Symes, David, and Kevin Crean. "Historic Prejudice and Invisible Boundaries: Dilemmas for the Development of the Common Fisheries Policy." In *The Peaceful Management of Transboundary Resources*, edited by G.H. Blake, et al. London: Graham & Trotman, 1995.

Symes, David, and Jeremy Phillipson. "Inshore fisheries management in the UK: Sea Fisheries Committees and the challenge of marine environmental management." *Marine Policy* 21 (1997): 207–224.

—— "Co-Governance in EU Fisheries: The Complexity and Diversity of Fishermen's Organisations in Denmark, Spain and the UK." In *Creative Governance. Opportunities for Fisheries in Europe*, edited by J. Kooiman, et al. Aldershot: Ashgate, 1999.

Tallberg, Jonas. "Responsabilité sans Pouvoir? The Agenda-Shaping Powers of the EU Council Presidency." Paper presented at the 4th Pan-European International Relations Conference at the University of Kent Canterbury, 8–10 September 2001.

—— "Delegation to Supranational Institutions: Why, How, and with What Consequences?" *West European Politics* 25 (2002): 23–46.

Thatcher, Mark, and Alec Stone Sweet. "Theory and Practice of Delegation to Non-Majoritarian Institutions." *West European Politics* 25 (2002): 1–22.

Thom, Mireille. "There is more to Compliance than Legitimacy and More to Policy than Institutions." In *Alternative Management Systems for Fisheries*, edited by D. Symes. Oxford: Fishing News Books, 1999.

Tollison, Robert D. and Thomas D. Willett. "An economic theory of mutually advantageous issue linkages in international negotiations." *International Organization* 33 (1979): 425–449.

Tsebelis, George *Nested Games. Rational Choice in Comparative Politics*. Berkeley: California University Press, 1990.

—— *Veto Players: How Political Institutions Work*. Princeton: Princeton University Press, 2002.

Tsebelis, George, and Geoffrey Garrett. "Legislative Politics in the European Union." *European Union Politics* 1 (2000): 1–36.

——"The Institutional Foundations of Intergovernmentalism and Supranationalism in the European Union." *International Organization* 55 (2001): 357–390.

Tullock, Gordon. "Public Choice in Practice." In *Collective Decision Making. Applications from Public Choice Theory*, edited by C. S. Russell. Baltimore: John Hopkins University Press, 1979.

——"Why so much stability?" *Public Choice* 37 (1981): 189–205.

—— *On Voting. A Public Choice Approach*. Cheltenham: Edward Elgar Publishing, 1998.

Wallace, Helen. "National Bulls in the Community China Shop: The Role of National Governments in Community Policy-Making.". In *Policy-Making in the European Communities*, edited by H. Wallace, et al. Chichester: John Wiley & Sons, 1978.

——"The Institutions of the EU: Experience and Experiments.". In *Policy-Making in the European Union*, edited by H. Wallace, and W. Wallace. 3rd ed. Oxford: Oxford University Press, 1998.

Wallace, Helen, and Fiona Hayes-Renshaw. "Collective Leadership, Confederal Bargaining and the Limits of Political Identity." http://www.ecsanet.org/conferences/5helen.htm, accessed on 07/19/2001.

Walton, Richard E., and Robert B. Mckersie. *A Behavioral Theory of Labor Negotiations*. New York: McGraw Hill, 1965.

Weber, Max. *Wirtschaft und Gesellschaft. Grundriss der verstehenden Soziologie*. Erster Halbband. Cologne: Kiepenheuer & Witsch, 1964.

Weber, Shlomo, and Hans Wiesmeth "Issue Linkage in the European Community." *Journal of Common Market Studies* 29 (1991): 255–267.

Weimer, David L. *The Political Economy of Property Rights. Institutional Change and credibility in the reform of centrally planned economies*. Cambridge: Cambridge University Press, 1997.

Wessels, Wolfgang. "Institutionen der Europäischen Union: Langzeittrends und Leitideen." In *Die Eigenart der Institutionen. Zum Profil politischer Institutionentheorie*, edited by G. Göhler. Baden-Baden: Nomos Verlag, 1994.

Westlake, Martin. *The Council of the European Union*. London: Cartermill Publishing, 1995.

Williamson, Oliver E. "The New Institutional Economics: Taking Stock, Looking Ahead." *Journal of Economic Literature* XXXVIII (2000): 595–613.

Wise, Mark. "Origenes y Evolución de la Politica Pesquera Comun de las Comunidades Europeas." *Revista de Estudios Agro-Sociales* 144 (1988): 9–35.

Zartman, I. William. *Ripe for Resolution*. New York: Oxford University Press, 1989.

Zürn, Michael. "We Can Do Much Better! Aber muß es auf amerikanisch sein?." *Zeitschrift für Internationale Beziehungen* 1 (1994): 91–114.

Index